OUR TURBULENT SUN

OUR TURBULENT SUN

Kendrick Frazier

PRENTICE-HALL, INC., Englewood Cliffs, N.J. 07632

Our Turbulent Sun, by Kendrick Frazier
Copyright © 1982 by Kendrick Frazier
All rights reserved. No part of this book may be
reproduced in any form or by any means, except
for the inclusion of brief quotations in a review,
without permission in writing from the publisher.
Address inquiries to Prentice-Hall, Inc.,
Englewood Cliffs, N.J. 07632
Printed in the United States of America
Prentice-Hall International, Inc., London
Prentice-Hall of Australia, Pty. Ltd., Sydney
Prentice-Hall of Canada, Ltd., Toronto
Prentice-Hall of India Private Ltd., New Delhi
Prentice-Hall of Japan, Inc., Tokyo
Prentice-Hall of Southeast Asia Pte. Ltd., Singapore
Whitehall Books Limited, Wellington, New Zealand

10   9   8   7   6   5   4   3   2   1

Library of Congress Cataloging in Publication Data
Frazier, Kendrick.
Our turbulent sun.

Bibliography: p.
Includes index.
1. Sun.   I. Title.
QB521.F73   523.7   81-5883
                    AACR2

ISBN 0-13-644500-4
ISBN 0-13-644492-X {PBK}

To F. E. (Jim) and Lenore Frazier,
my parents,
who have always
encouraged my curiosity

CONTENTS

FOREWORD

At night we see them all around us, burning in the black of space. We are surrounded by stars: above, below us, reaching out in all directions, farther than the eye can see. A hundred billion stars make up our Galaxy and each is much like all the others—distant fires that we shall never know. One star, our Sun, is special. Of all the billions it alone is close enough to see in any detail, and near enough to have any real effect upon our lives.

But what an effect! Our dependence on this one star for light and heat, and food, and air, is utter and complete. Without it there would be no life. And little poetry: without the Sun there would be no clouds or rain, no bright blue sky or running stream, no rainbow, no seasons, and no day. One star: so important to us, yet so little understood.

It seems to me important to learn all we can about our Sun, and surely the most practical of all questions is its constancy. To almost all who have lived under its benign grace, the Sun is the very symbol of constancy. Daily it rises and daily it sets, looking very much the same. More quantitatively, from fifty years of mountaintop measurements astronomers have shown that its total output changes little: probably less than one percent in all that time. Only recently, from spacecraft above the atmosphere, have we been able to measure any variation at all in the Sun's intrinsic brightness; the greatest change yet recorded with these sophisticated instruments was a flicker of two-tenths of one percent, in early April of 1980: a springtime murmur so soft and subtle that no one could have sensed it on the ground. Moreover, fossil records from the Earth and rock surfaces on the Moon both attest that on average, the Sun of several billion years ago was much like the one we see and feel today.

Those who study the surface physics of the Sun will tell a

somewhat different story. The occurrence of sunspots and their day-to-day changes leave no possible conclusion but that the Sun is imperfect, inconstant, irregular, and a magnetically-variable star. Explosive solar flares remind us that changes on the Sun are violent as well, with impacts that can be felt on Earth, nearly 100 million miles away. Solar physicists also know that our direct knowledge of the temporal behavior of the Sun is very limited: we have monitored the Sun in detail for but a moment of its life. Is it prudent to assume that in a hundred years, or two hundred, we have witnessed all that it can do? Indeed, historical accounts and tree-ring isotopes reveal past episodes of solar behavior that would concern and surprise us were the Sun to repeat them today. Several of these long spans of misbehavior fall within the scope of modern history, and there is some evidence that each of these known, major changes was accompanied by changes in the climate of the Earth.

Thus an accurate description of the Sun falls somewhere between extremes of chaos and constancy, though surely nearer the latter. Recent work from a number of angles and by many careful scientists has emphasized a changing Sun, and indeed our study of the star is now in a time of ordered readjustment. It is this story that Kendrick Frazier treats so carefully and well in the pages that follow.

There is much left to learn. The challenge of modern solar physics is to determine the real extent of changes on and in the Sun, on all time scales, and eventually to trace each burst or solar flicker through miles of space and the convoluted filter of our atmosphere to find its end effect upon the Earth. It may well be that solar changes are so slow, or subtle, to be of academic interest only. Even so, all that we learn about the nearest star—its sources of energy, internal structure, constancy or inconstancy—can be applied to the hosts of other stars and thus to all the cosmos. "What is the Sun?" asked astronomer Amédée Guillemin in 1870: "If the science of astronomy could solve this great problem, it would be nearly capable of solving that of the entire Universe."

Dr. John A. Eddy
High Altitude Observatory
Boulder, Colorado

ACKNOWLEDGMENTS

Our sun is only one of a hundred billion suns in our galaxy, and our galaxy only one of perhaps a hundred billion such star systems in the universe, but to us it is unique: It is life itself. This is a book about the dramatic scientific quest for a new understanding of our sun, with an emphasis on solar variability, the recent revelations of surprising irregularities in our own star. It is about the effects events on the sun—some of colossal scale and force, some of exquisite subtlety—may have on earth. And it is about some of the people, past and present, who have been pioneers in the study of the sun's variations and effects. For science is a very human activity, and I think it is more honest and interesting to present it that way. Although much—perhaps most—of what is related here is new in the past five years or so, I have tried wherever possible to place the discoveries and issues into some kind of historical context. Astronomy has a wonderfully rich history, and I've drawn from it with pleasure and an appreciation of both human foibles and human creativity. I have tried not to shy from controversy. Much of the subject matter here is still unsettled. That is typical of any area of science in vigorous intellectual ferment. The quest for a solar connection to weather and climate variations on earth—the subject of my chapters 8 through 12—is one in which disputation is perhaps the primary characteristic. It is, in fact, one of the epic scientific controversies of our time. I have tried to present an honest appraisal of the current status of the sun/climate question, but I recognize that each participant and informed observer will have his or her own view.

I have drawn on the knowledge and assistance of a number of scientists. I owe debts to many. John A. Eddy of the High Altitude Observatory was through his work a source of stimulation and through his words a source of encouragement and support, before, during, and after the writing. I especially thank also Walter Orr Roberts and Roger Olson of the Aspen Institute for Humanistic Studies, Henry A. Hill of the University of Arizona, John M. Wilcox and Philip H. Scherrer of Stanford University, Richard C. Willson of Jet Propulsion Laboratory, and Gary Hechman and Howard H. Sargent of the Space Environment Services Center. They all gave of their time in interviews, read portions of my draft manuscript related to their work and expertise, and offered valuable suggestions for improvement. I am also grateful to S.-I. Akasofu of the University of Alaska and Ralph Markson of M.I.T. for similar assistance. Although I have drawn on their help to ensure accuracy and completeness, all responsibility for the content is mine. I also owe thanks to Dimitri Mihalis, Robbin Stebbins, and their colleagues at Sacramento Peak Observatory, John Imbrie of Brown University, Minze Stuiver of the University of Washington, and Gordon Newkirk of the High Altitude Observatory. Thanks also to Oscar Collier, Hugh Rawson, Diane Johnson, Elizabeth Maggio, and Sue Lyons, and in ways too special to enumerate, my own family, Ruth, Chris, and Michele.

1. A SUN FULL OF SURPRISES

"Is the Sun Still Shining?" "What's Wrong with the Sun?" "Is the Sun Shrinking?" "The Inconstant Sun."

You might think these headlines were from those tabloid newspapers that stare flamboyantly at you from the supermarket checkout counter. Not so. These titles appeared over reports in some of the world's most respected scientific publications in the past few years.

They reflect a genuine concern and excitement in the scientific community over a number of recent revelations about the sun. These discoveries have shaken scientists' calm assumptions about a regular, dependable sun that functioned smoothly like clockwork. It is now apparent that beneath the sun's seemingly uniform yellow-white surface that most of us take so much for granted is "a riddle wrapped in a mystery inside an enigma," to borrow Churchill's famous line.

Who would think that what the ancients considered a symbol of the absolute perfection of God and that what we today consider merely an ever-present, reliable source of light and energy could pose so many fundamental and troubling questions to late-twentieth-century science?

This is a book about those puzzles. A few of the long-standing riddles about the sun have been satisfactorily solved in the past five years, only to be quickly replaced by others. Solar science is in a state of ferment. It is a refreshing time for inquirers into the workings of nature when old ideas are challenged and new ideas sail hopefully forth, often to go quickly aground on the rocky shoals of critical scrutiny. But some will survive the journey and help shape the still-evolving conception of the changing sun.

I find it exciting, and I want to share with you some of the sense of adventure that study of the sun today brings.

Why is the quest important? We live in the nourishing river of the sun's star streams, warmed, lighted, fed, fueled by its emanations— our origin, our continued existence deriving from its presence. No wonder that from the earliest times till today, the sun has been revered and held in awe, correctly seen as a giver of life. No wonder its seasonal wanderings in the sky were tracked with precision. To the Egyptians the sun was the god Ra, to the Greeks, Helios. To the modern-day Hopi, who call it Tawa, it is revered not only for its generative powers but as the one who "keeps the ways," providing structure and order to all life. Our modern age, grown more conscious of its debt to and inextricable links with nature, has honored it with a Sun Day and seeks to find economical ways to tap its power as a future source of clean and abundant energy for humankind. No matter whether it be spiritual reverence or secular celebration, whether the motivating force be mystical or practical, we feel toward the sun a bond, a familial warmth and relationship appropriate for no other heavenly object. Any significant unexpected change in the sun's behavior can affect us in profound ways, both psychologically and physically.

Three hundred seventy years ago when curious Europeans turned the first telescopes on the sun and discovered slowly changing features on its surface, the revelation was a shock to the theological image of the sun as a timeless, unchanging entity. Something akin to that shock is what solar astronomers have been experiencing in the past five years. The concept of a steady sun has been shattered, and with it the sense of stability and assurance that our conception of solar regularity brought to our lives.

This is troubling, yes, for all shifts from long-held comforting conceptions are troubling. Yet it is exciting too, for it has brought new vigor to the science of the sun and it holds the promise of opening new pathways to understanding some basic ways in which nature works. And that, after all, is what the process of intellectual inquiry we call science is all about. Who knows what new insights may come out of it? As solar scientist E. N. Parker has said, "Solar physics is the mother of astrophysics." The sun has been the origin of most of our concepts about the processes going on inside other stars and galaxies. Those concepts have shaped our understanding of the cosmos and our place in it.

And who knows what practical gains may in turn come from new insights into the sun? The sun is a natural laboratory in space for the study of a state of matter called plasma, and our scientists are searching for ways to mimic the sun's thermonuclear generation and confinement of high-temperature plasma as a means of producing energy from the atoms of seawater. Anything we learn about how the sun works, or doesn't work, may have immense consequences for our future.

For now all we can say is that we appear to have taken the sun for granted All is not as it once seemed. As the solar physicist Gordon Newkirk recently said, "Several icons of solar physics have fallen rather noisily." The sun is not now necessarily as it has been in the past, nor as it will be in the future:

● *It seems to be subject to long irregularities and interruptions in its cycles of activity.*

● *It seems to fluctuate slightly in luminosity over short periods of time. The "solar constant" may not be constant.*

● *It is now also seen to pulsate or vibrate, with periods ranging from minutes to hours.*

● *Recently there has even been concern that it may be shrinking, at least temporarily.*

● *All these changes and uncertainties are of considerable interest in themselves, but they also have ramifications for the long debate over how variations in activity on the sun may affect the earth and our fluctuations of weather and climate. That is a controversial and important subject in itself, one we'll take up in the latter part of the book.*

What is this sun, this fiery object of our affection and concern? Anaxagoras proposed in about 434 B.C. that the sun was a mass of fiery stone and calculated that it was 4,000 miles above the surface of the earth and as large as the southern Greek peninsula, about 35 miles. He was arrested for disputing the established religious dogma.

In the third century B.C. Aristarchus of Samos, who far ahead of his time considered the earth a rotating sphere which along with the other planets revolved around the sun, did some scientifically well-reasoned geometric calculations. He judged the sun to be nineteen times farther away than the moon (whose distance he underestimated), or a distance of some 1,150,000 kilometers (720,000 miles). Had his instruments not been so crude he might have come close to the right answer. Still he recognized that the sun was at least as large as the earth. In the second century B.C. Hipparchus used his observations of a lunar eclipse to accurately calculate the distance to the moon. He then used Aristarchus's factor of 19 to get a new limit for the size and distance of the sun. The result was to show that the sun's diameter was at least seven times as large as the earth's, a quite remarkable realization! That figure remained unchanged for 1,700 years until the seventeenth century A.D. Finally, in 1670, the Italian-French astronomer Giovanni Cassini did a new calculation that was within 10 percent of the proper value.

We now know the sun's vital statistics quite well. We know it is at an average distance of a little less than 150 million kilometers (93 million miles), or 389 times the earth-moon distance, and is a vast sphere of gas about 1,400,000 kilometers (865,000 miles) across. More than 109 planets the size of earth would fit across its width.

How can we better visualize the vast size and distance of the sun? I've done some figuring. If the sun were reduced to a spherical dome exactly covering the width of midtown Manhattan from the Hudson to the East River (such a dome would enclose four Empire State Buildings stacked on top of each other) the earth would be a sphere 30 meters (100 feet) wide halfway between Cleveland and Detroit. Or, if the earth were reduced to the size of a golf ball, the sun would be a sphere more than 15 feet in diameter (the volume of a good-sized room) more than six tenths of a mile away. Or if the earth were reduced to the size of the period at the end of this sentence, the sun would be half the width of a line of type on this page and 11.7 meters (38 feet) away.

The sun is so voluminous it could hold 1.3 million earths, but since its mean density is only about a fourth of earth's, its total mass is about 330,000 times that of earth. If you want a tonnage figure it would be the number 219 followed by 25 zeroes. Or, in kilograms, 2 followed by 30 zeroes.

In fact, the sun's mass is 745 times that of all the other planets of the solar system put together, including giant Jupiter. It is clearly the dominant force in the solar system, even without considering its output of energy.

It is, of course, a glowing ball of gas, 90 percent hydrogen, 9 percent helium, 1 percent other elements, powered by nuclear fusions in its hot (10 to 16 million degrees C), dense central core. Four hydrogen atoms fuse to become a helium atom in a three-step proton-proton reaction. The resulting helium nucleus has a slightly less mass (0.7 percent) than the original four hydrogen atoms. It is the conversion of that missing mass to energy that powers the sun and lights our sky. The process is about a million times more energetic than a chemical reaction, such as burning. Our weapons makers have mimicked it in the design of the hydrogen bomb.

Although fusion itself happens quickly, the practical release of energy to the sun's surface is a slow process. The solar interior is so opaque to radiation that the energy takes somewhere from 1 million to 50 million years (estimates vary) to work its way out to the solar surface. Once it gets out to the surface, the light takes only a little over 8 minutes to reach the earth. As we shall see, energetic particles spewed out by the sun travel at far less than the speed of light and reach earth in several days. I find it interesting to contemplate that the long fight outward from the center to the surface of the sun means that the light we are now receiving had its origin in events initiated in the solar interior before the dawn of human history.

The surface of the sun is far less hot than the interior, about 5,700°C, or less than 11,000°F.

Since the sun is a sphere of gases and not a rigid solid, different sections of it can rotate at different speeds. That is exactly what happens. As seen from earth, the sun's equator completes a rotation in less than 27 days. But the rotation slows poleward. One third of the way toward the poles the rotation rate exceeds 28 days; two thirds of the way, 30 days.

As stars go, the sun is quite average in type, size, and age. But that fact hardly diminishes its paramount importance to us. There may be other stars larger or smaller, brighter and dimmer, older and younger. But they are all hundreds of thousands to billions of times farther away from us than the sun (even considering only those stars in our own galaxy). To us, their radiation is as inconsequential as the sun's is all-surpassing. And besides, as we shall see, the sun is a star of immense interest and mystery in itself.

The conception of a changing sun would have been alien and repugnant to the preeminent natural philosopher of Greek times, Aristotle. Aristotle taught that the sun was a perfect body without blemish. This view then became a belief of Christian orthodoxy during the Middle Ages. This is very likely why there are so few records of observations of features on the sun during the first fifteen centuries of the Christian era. After all, why look for changes on a body that is known to be perfect and unchanging?

No such constraints retarded solar observation in the East. Chinese astronomers recorded well over a hundred naked-eye observations of the features we now call sunspots during the period 28 B.C. to A.D. 1638. The histories of Japan and Korea contain references to sunspots too.

Sunspots are extremely difficult to see with the unaided eye, however. It's not that they're small. Many are quite large enough. It's that the intense glare makes it difficult (and quite dangerous) to look at the sun directly. They're occasionally perceptible to the naked eye when viewing the sun through a thick haze.

The invention of the telescope changed all that. With a telescope the image of the sun can be not only enlarged but projected onto a screen. No fewer than four men are credited with the nearly simultaneous first telescopic observation of sunspots in 1611. Priority of first publication goes to a twenty-four-year-old German observer, Johannes Goldsmid, better known as Fabricius. He was the son of a theologian who took a special interest in astronomy. Next was Galileo Galilei, in Italy. The third was Christoph Scheiner, a German-born Jesuit priest and mathematician, later a professor at the University of Rome. And the fourth was Thomas Harriot, in England, a onetime mathematical tutor to Sir Walter Raleigh who had also for a time worked as a surveyor of the New World in Virginia. He was then doing scientific work under the patronage of the Earl of Northumberland.

Scheiner at first thought the sunspots must be either a flaw in his telescope or small planets revolving around the sun. Galileo, however, quickly realized they were indeed features of the sun itself. They changed their size and shape. Not only that, but he inferred from them the sun's rotation period, and showed that sunspots occurred in groups and in two bands above and below the sun's equator. Scheiner, after receiving three letters from Galileo detailing his observation, had to agree that the spots were on the sun itself, although a long and bitter controversy ensued between the two men over their personal claims to the discovery.

The hazards of announcing unsettling discoveries about nature weren't to be taken lightly in those times. When Scheiner reported his discovery to his ecclesiastical superiors, they refused to believe it. He was not allowed to publish the news under his own name. Galileo, too, delayed publication. There was risk in sticking one's neck out too far.

"And I, indeed, must be more cautious and circumspect than most other people in pronouncing upon anything new," Galileo wrote to the friend who brought Galileo and Scheiner into communication. "As Your Excellency well knows, certain recent discoveries that depart from common and popular opinion have been noisily denied and impugned, obliging me to hide in silence every new idea of mine until I have more than proved it."

That was only five years before Galileo was summoned on order of the Pope and warned neither to defend nor to hold the heretical views about the earth going around the sun advanced by the astronomer Copernicus. And in 1633, in the Hall of the Inquisition, Galileo was tried and forced to kneel and publicly recant his heresies. He was pointedly shown the Inquisition's instruments of torture and placed under permanent house arrest. All in all, he was perhaps lucky. In 1600, Giordano Bruno, for advancing the same views, had been burned at the stake.*

Sunspots weren't as threatening as the Copernican world view. Nevertheless, further important discoveries about changes on the sun didn't come until the nineteenth century. (That may have been due largely to a significant shortage of sunspots during much of the intervening time, a subject that we will get to.) The 1800s brought us a conception of the cyclic sun.

A German pharmacist named Heinrich Schwabe, whose hobby was astronomy, was responsible. Schwabe was obviously a patient, methodical man. The discovery he is now remembered for came as a result of a long period of study having an entirely different

---

*In October 1980 the Vatican announced it would reopen Galileo's case. The wheels of justice sometimes move with celestial slowness!

intention, a not altogether unknown situation in science. Schwabe
wanted to discover any new planets that might exist inside the orbit of
Mercury. To that end he began in 1826 charting all spots on the
surface of the sun each day because any undiscovered inner planet
might show up as a small dark disk when it passed in front of the sun.

In 1843, Schwabe, then fifty-four years old, announced that
he had detected a cycle in the occurrence of sunspots. They slowly
became more abundant and then less so in a cycle of ten years (now
known to be eleven years). What was the reaction? None. Practically
no one heard of his find. It wasn't until 1851, when the famous German
naturalist Alexander von Humboldt published a table of Schwabe's by
then twenty-five years of sunspot statistics, that the world took note.*
Schwabe had confirmed something that other astronomers had failed
to notice in more than two centuries of sunspot observations. Their
occurrence was cyclic. The Royal Astronomical Society awarded him
its gold medal for the achievement.

I can understand the hold that sunspot observations can
have on a person. I have a small reflecting telescope set up in my
office, an arm's length to the left of the desk where I type. It has a clear
view of the sky to the west out a sliding glass door, and any afternoon I
can quickly line it up with the sun and project onto its screen the
sunspot-flecked solar image.

Suddenly a solar surface that to our eyes is only a brilliant,
flat, blank disk, too painful even to look at, becomes transformed into
something else altogether. It's like a distant planet with its own mobile
geography, a bewildering array of features spread across its breadth.
And the sun itself is no longer two-dimensional but full-bodied, its
sphericity impressing itself on our consciousness like a sculpture
suddenly bursting forth from the bounds of its clay.

Each day there's something different. Most of the sunspots
from the previous day, some large, many small, are still there. But
some have grown and become organized into clusters. Others have
begun to deteriorate. The range of sizes is extraordinary, from barely
visible specks to vast assemblages covering an area perhaps a tenth
the diameter of the sun's disk.

And their structure is clearly visible, even with my small

---

*Humboldt published it in his five-volume work Kosmos, *a highly popular account of the
earth in cosmic perspective and of the joy of scientific discovery. A worthy forerunner, by
130 years, to Carl Sagan's* Cosmos!

telescope. Each has a dark central area, which astronomers call the umbra, surrounded by a lighter area, the penumbra. Thin lines or filaments radiating out from the umbra give the penumbra a delicate texture. Sometimes the central region is only a small portion of the total area. Other times it's almost half. Sometimes the umbra is broken up into two or more sections of different sizes. The sunspots certainly do occur in groups, often in long lines, like chains of oceanic islands.

They are every conceivable shape: round, oval, oblong, sickled, teardrop, and everything in between. One I am looking at right now, for instance, has a lumpy penumbra with a black umbra in the shape of a thick Y. Another, smaller one in the southern hemisphere of the sun has more the shape of a tadpole. And sometimes there are associated structures trailing out behind them.

They do occur in two parallel bands above and below the sun's equator (I can confirm Galileo). It is easy to notice the sun's rotation—something one otherwise has no visual or intuitive inkling of—by watching the groups progress across the disk from left to right, day by day. As old ones disappear off the right edge, after nearly two weeks onstage, new ones appear to the left. At first they're distorted and indistinct because you're seeing them from such a slant, but in a couple of days their true structure becomes easily apparent.

I charted the sun's surface this way for almost two months near the height of the current sunspot cycle. It's exciting each day as you bring the image into focus to see the changes that have taken place in twenty-four hours. If cloudy weather interrupts for a day or two, the disk can be almost unrecognizable when you again get to view it. Two days can bring enormous change. But particularly large and distinct features can retain their mark for many days, even weeks.

One assemblage first appeared as a pair of equal-sized sunspots, side by side, rather than in tandem, each with a distinctive thin dark tail, followed by another large spot and some smaller ones. This group was also preceded by several large, distinctive spots, the whole business extending over a third of the sun's diameter. Over the next five days, the two spots at the head merged to form one very large one, and the tails widened into two broad diverging bands, like a man's legs. In another two days, the head had broken off from these bands to form one enormous sunspot with a large dark umbra, and the trailing legs were beginning to break up. As the sunspot at the head approached the far edge of the solar disk, it took on a decidedly elongated shape, a result of the changing perspective, and disappeared from view, trailing the now fragmented and indistinct parallel bands that had accompanied it for nearly two weeks.

I was hoping some sign of this remarkable group would stay intact long enough to make an encore appearance at the sun's left edge in two more weeks. But when the time came, the show wasn't the same. There were several spots, but the distinctive arrangement was not there. I couldn't be positive whether any of them were the ones I had followed before. I wanted to think so, but I had to admit I couldn't be sure.

A warning to any would-be sunspot watchers: *Never—I repeat, never—try to look at the sun directly through a telescope. That way lies blindness. Always project the image onto a screen attachment. Just one second of concentrated sunlight on the retina can destroy your vision.* I use a sun diagonal prism, which reflects only 10 percent of the heat and light and projects an upright image of the sun downward onto a horizontal screen.

What exactly are these sunspots? We know they appear darker than the rest of the sun only because they are about 2,000° C cooler. Even that is misleading. They are actually as bright as a carbon arc light. It is only the contrast with the rest of the still brighter solar surface that makes them appear dark.

Using special instruments he designed, the American astronomer George Ellery Hale discovered in 1908 that sunspots are areas of intense magnetic fields. Their strengths are thousands of times greater than the general magnetic fields of the sun and the earth. The sunspots often occur in pairs, the two spots having opposite magnetic polarities, like the north and south ends of a magnet.

We don't really know how they get amplified to that strength, although we do believe magnetic fields on the sun are a product of the sun's differential rotation acting on the electrically charged gases churning through the outer one fourth of the sun known as the convective zone.

One theory about the origin of sunspots starts with the idea that north-south lines of magnetic field extending from one pole of the sun to the other become wrapped around the sun in an east-west direction, due to the sun's rotation. These twisted lines of force lie generally beneath the visible surface in the form of tubes. But where one of these tubes rises up through the surface the plane of intersection becomes a sunspot. Where the magnetic tube loops back down beneath the surface again the spot of opposite polarity is formed. Their concentrated magnetic fields inhibit the transport of energy upward to the visible solar surface and thus the spots are cooler and darker.

So sunspots are intimately related to the forces that create strong magnetic fields on the sun. Since the time of Schwabe we've known that their abundance rises and falls over an 11-year period, the so-called 11-year sunspot cycle (actually 11.2 years on average). We now also know that at the end of one sunspot cycle, the sun's general magnetic field gradually reverses polarity. The north magnetic pole becomes the south magnetic pole, and vice versa. So the sun also demonstrates a 22-year magnetic cycle, or double sunspot cycle. It truly is a cyclic sun.

But are these the cycles we are talking about when we say the sun is changing in ways we previously unsuspected? After all, we've known about sunspots for many centuries and about the sunspot cycle for 130 years. And we've been aware of the 22-year magnetic cycle since the 1950s.

No. What's new is a series of realizations about ways the sun changes and fluctuates in addition to the cycles of sunspots. Sunspots were the first sign that the sun wasn't the perfect unchanging sphere of antiquity. And since they are so readily visible they have always been given great attention. But it appears that the seemingly regular recurrence of the sunspot cycles may have deceived us into thinking of the sun as a smoothly running clock, always functioning just as expected. Now that view is changing. The sunspot cycles themselves, as we shall see, are not so dependable as we thought. That fact alone has started to have enormous revisionary consequences for our conception of the sun. Beyond that, many of the events in the sun that are most likely to affect us directly may have little to do with sunspots. The sunspot cycles seem to have been somewhat overrated as a convenient marker of solar activity.

Other revelations about variabilities of this star of ours have come from observatories out in space and detectors far underground, from the musty pages of past historical records and from still-continuing modern measurements of unprecedented precision. We are learning that we live in the streams of a star whose emanations and behavior may change in ways we never suspected. It seems to be an uncertain sun we are living with. Certainly it is a sun full of surprises.

2. THE SUN'S IMPACT ON EARTH

The sun offered no sign of what was to come. The area soon to be memorialized as solar active region 331 had been seen at birth three weeks earlier during the previous solar rotation and tracked till it passed beyond the edge of the sun's disk. During that period it had grown only moderately and discharged no flares of significance.

The solar season gave no reason for suspicion either. The entire solar cycle beginning eight years earlier had been disappointingly uneventful. The period of apparent maximum solar activity had come and gone three and a half years earlier with only a few large events to show for it. The solar cycle was now well along in its declining phase. The expected minimum was only three years away.

The sun's rotation would bring region 331 back into view again on July 29. No one expected it would be anything but an insignificant and declining feature. Its reappearance on schedule on the northeast edge of the sun's disk was at first uneventful. But the innocent cluster of spots that had disappeared from view 14 days earlier had grown tremendously. It had become organized. A large pair of sunspots had formed. The structure of the whole region was complex. There were signs of instability. By the next day, on July 30, Mount Wilson Observatory in California had analyzed its magnetic characteristics and confirmed that the region was complex. Energetic emanations seemed possible.

Solar forecasters at the Space Environment Services Center, the solar "weather bureau" in Boulder, Colorado, monitored the region closely. They recognized signs that new magnetic fields were emerging from beneath the sun's surface. Their interaction with the already present magnetic fields can cause the development of even greater magnetic complexity, which could trigger significant solar events. The scientists issued a forecast for probably sporadic but intense solar flares.

On August 2, active region 331 cut loose. A series of three energetic flares blasted off the sun. The first and third, fifteen hours apart, were the most powerful.

A large flare is a catastrophic explosion on the sun. Energized by conversion of powerful, pent-up magnetic forces, it can release the energy of 10 million hydrogen bombs in a few minutes' time. One significant measure of solar-flare intensity is the energy of its X-ray emanations. The categories run from (starting at the lowest intensity) C1 to C9, M1 to M9, and then X1 to X9 (the greatest). The first and third flares were rated X2. Their optical outputs were also high. Their radio emanations were several thousand times more energetic than the sun's normal levels. And an increased rain of protons, atomic particles from the sun, was beginning to be monitored at earth. The sun was about to get the world's attention. The Boulder solar forecast center issued this announcement.

Solar activity has been high. Region 331 has produced a series of energetic flares.... A proton event...and magnetic storm, all of at least moderate intensity, are expected from the current activity. Proton flux should rise sharply in the next 12 hours. The geomagnetic storm should be most intense on 04 and 05 August. The probability for further energetic flares from the sun is unknown.

The worst was yet to come. August 3 brought a temporary lull in the surge of solar activity. Only two flares issued forth from 331, both small. But the effects of the emanations of the previous day were still radiating outward from the sun.

The United States' Pioneer 9 spacecraft was orbiting the sun between Venus and earth. It was thus closer than us to the sun by about 35 million kilometers (20 million miles). At 11:24 universal time,*

To avoid the confusion of time zones, astronomers use the time of the Greenwich (England) time zone, which they call universal time (UT). You can subtract 4 hours to get Eastern standard time, 5 hours for Central, 6 for Mountain, and 7 for Pacific standard time.

it felt and reported an interplanetary shock wave from the first flare the previous day, 33 hours earlier. The solar wind, the stream of charged particles continually emanating from the sun, had suddenly increased in speed from 350 to 585 kilometers per second (780,000 to 1,300,000 miles per hour). Later a shock front from the second flare would see the solar wind increase to an incredible 1,200 kilometers per second (2,660,000 miles per hour).

Pioneer's advance warning allowed the solar forecasters in Boulder to issue an explicit warning that during the night of August 3–4 the earth was to experience the sudden onset of the first of several geomagnetic storms.

This shock wave should reach the earth and produce a sudden commencement at about 0030 04 August. The storm main phase and associated auroral activity is expected to begin about 0700. A second sudden commencement from the 02/2000Z flare* is expected later on the 4th. The combined effects of these two flares should produce a magnetic storm of minor intensity on the 4th and major intensity on the 5th.

A geomagnetic storm is like no other kind of storm on earth. It is a disturbance not of systems of air masses and weather but of the earth's magnetic field. And it is a worldwide disturbance, not a local one.

The earth's magnetic field is the result of the dynamo action of electrical currents generated in the slowly moving molten conducting iron in the planet's fluid core. The effect is much as if a giant bar magnet were at the earth's center slightly inclined to the axis. It is, of course, what makes a compass needle point to magnetic north.

A bar magnet has lines of magnetic field looping far out from all points on its surface. Iron filings on a sheet of paper placed over a magnet reveal these symmetrical curves of its magnetic field. In the same way, the earth's magnetic-field lines curve far out into space and, except for the polar regions, return to the surface at some other point on the planet in a series of grand symmetrical whorls.

The solar wind of charged particles from the sun is continually flowing around and interacting with earth's magnetic field. But a giant flare on the sun sends vastly greater quantities of these particles at accelerated velocities at the earth. When they, and the solar magnetic fields they drag with them, come into contact with our planet's

---

*The third flare of the previous day, August 2, at 2000 UT, also sometimes called Zulu time (Z).

magnetic field, they give it a silent but significant twang. Together with the flare's ultraviolet rays, X rays and radio waves, they disturb the earth on a grand scale.

The most visible manifestations of geomagnetic storms are the beautiful auroras, or polar lights, which have charmed and terrified societies for thousands of years. But they also induce voltages in the earth's upper atmosphere, in power cables, and in the ground. And those voltage changes can alter or interrupt the current in long-distance electric power transmission lines. They can damage electronic equipment. They can electrify pipelines. They can even affect networks of radar sentries set up to detect the first signs of enemy attack.

The disturbances can create a temporary new layer in the earth's ionosphere that absorbs shortwave radio transmissions, blacking out important commercial and military messages. Radio transmissions in the arctic and antarctic can be totally blocked for hours. All these communications effects were soon to be felt.

Because it results from the arrival of a surge of particles from the sun, the onset of a geomagnetic storm is nearly always sudden. The August 4 storm struck just as the forecasters had predicted.

At Boston College, the Weston Observatory magnetometer,

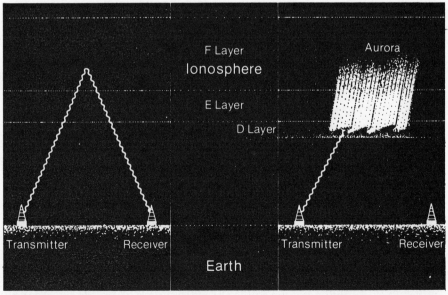

Geomagnetic storms can cause intense auroral displays. Often an abnormal layer of the ionosphere, the D layer, is formed, absorbing radio waves (S.-I. Akasofu/Alaska Geographic)

an instrument for measuring the intensity of the geomagnetic field, recorded an abrupt increase at 0118. One minute later it was registered by the magnetometer at the Boulder solar forecast center. Forty-six hours had elapsed since the eruption on the sun of the flare that caused it. At Boulder, the disturbance would continue for nineteen and a half hours, after which a second storm, caused by the other major flare on August 2, would begin.

At 0400 the aurora was simultaneously sighted from Illinois, Wyoming, and Colorado. At Hanover, Illinois, observers described it as a broad-banded, brick-red structure that began in the west and slowly moved eastward. It lasted for two hours. In Savory, Wyoming, it was seen as a red glow in the northwest, lasting thirty minutes. In Westcliff, Colorado, it was a patch of red high overhead preceded and followed by a red arc low in the north. In Hillsboro, Oregon, it was seen as a strong white steady glow in the north-northeast, topped by intermittent narrow bands.

Just about this time, at 0621, another giant solar flare blasted loose from region 331 on the sun. First to photograph it was the solar-patrol camera in Athens, Greece. X-ray observations later that day determined that it rated a category of at least X5, the biggest so far of the four in the series. It was actually more intense than that, but the satellite sensors became saturated at that point and could give no higher reading. Its main effects would be felt on earth in three days.

In the meantime, a full-scale geomagnetic storm was now raging over the earth. The alert from Boulder was succinct.

ADALERT BOULDER 04/0930Z
STRONG MAGSTORM IN PROGRESS

Throughout August 4 and 5 the storm continued at an intensity that scientists rated in the rare category "Great Storm." By now the effects of the second flare of August 2 had also reached earth. The consequences were rapidly spreading.

At 2230 UT, late afternoon in the American Midwest, a solar-induced direct current of 60 volts was measured on an American Telephone & Telegraph underground coaxial telephone cable between Chicago and Nebraska. A few minutes later, the Bureau of Reclamation power station in Watertown, South Dakota, experienced voltage fluctuations. One transmission line normally carrying 230,000 volts had voltages as high as 235,000 and as low as 205,000 volts. At Northern States Power in Minneapolis, a protective circuit breaker shut down a transformer. Two minutes later, Wisconsin Power and Light, Madison Gas and Electric, and the Wisconsin Public Service Corporation were all experiencing voltage fluctuations on their lines.

At Deer Lake, Newfoundland, solar-induced ground currents activated a protective device on a 25-megawatt converter belonging to the Bowater Power Company. It was the first of many times this would happen over the next day. At times solar-induced currents as high as 200 amps were coursing through the ground in Newfoundland.

As nightfall came auroras danced across the skies of the northern hemisphere. Observers witnessed the aurora in France, Czechoslovakia, Switzerland, and Great Britain. In southern England its structured red glow was seen straight overhead. Darkness in North America brought auroral displays to Canada and the Upper Great Plains states. From Winnipeg, the aurora was visible nearly to the southern horizon. From Sioux Falls, South Dakota, it covered the entire northern half of the sky.

The night brought frequency fluctuations to the power system of the Idaho Power Company. The Bowater Power Company's protective relays were tripped five more times. A 230,000-volt transformer belonging to the British Columbia Hydro & Power Authority exploded. Manitoba Hydro in Canada found that the flow of power it supplied to Minnesota dropped from 164 megawatts to 44 megawatts for about a minute, and later from 105 megawatts to an average of 60 megawatts for ten to fifteen minutes.

By now the solar storm had become an international news story. The headlines captured the drama:

"The Sun Erupts." "Storm on the Sun." "Earth Swept by Magnetic Disturbances." "Solar Explosions Felt on Earth." "Intense Magnetic Storms Hit Earth." "Solar Flares Peril Electric Power." "Solar Flares Disrupt Earth." "The Sun Puts on a Dazzling Show."

Australian authorities reported that all sections of the transpacific coaxial cable experienced voltage variations. Fortunately, all power and communications lines in Australia remained unaffected.

Canada wasn't so lucky. Both the Canadian Overseas Telecommunications Corporation and Canadian National Telecommunications found that the large voltage variations from the August 4 and 5 magnetic storms had damaged filters and burned carbon blocks in their systems.

Ships navigating the St. Lawrence River had their high-frequency radio reception interrupted. Navigation beacons experienced erratic phase shifts. Compasses temporarily deviated from pointing to magnetic north.

Active region 331 let go with still another giant flare on August 7, this one at least the equal of the other powerful ones. The earth was in for several more days of magnetic storms. Region 331 would soon fortunately be disappearing around the edge of the sun. Even as it did so on August 11 it ejected another flare, photographed in profile as a giant loop prominence, and satellites with a view of the part of the sun facing away from the earth noted that region 331 kept popping for several more days.

It wasn't until 331's last effects had faded away on August 13 that the Boulder solar forecast center could issue the following forecast:

THE SOLAR DISK IS STABLE WITH ONLY FOUR SMALL SPOT GROUPS VISIBLE.. . QUIET GEOMAGNETIC CONDITIONS ARE EXPECTED.

One can almost hear the sigh of relief.

The great solar flares of August 2–7, 1972, dramatize some of the ways the sun's actions can affect earth and its people. The solar flares in that five-day period and the resulting geomagnetic storms are still the most energetic observed since the Space Environment Systems Center began its continuous twenty-four-hour watch on the sun in the mid-1960s. But this story could just as well be set in the future, for the same kinds of happenings could occur again soon.

They vividly demonstrate that without a doubt the sun is quite capable of unleashing major events during most phases of its eleven-year cycle, not just at its peak. As a later report of the events stated, "They showed that geophysical events of historic proportions can occur in a solar cycle having a history of relatively low activity." In other words, there is practically no time that the sun cannot erupt, raising havoc with our technological systems on earth.

The events of August 1972 were better documented than any previously. The immediate and widespread attention they drew from the public was also unprecedented. But it would be misleading to suggest that they were unique. All of the phenomena and effects observed had been reported many times in the past.

In fact, it turns out that even in this century, the intensity of the 1972 storms has been exceeded many times. On an index of

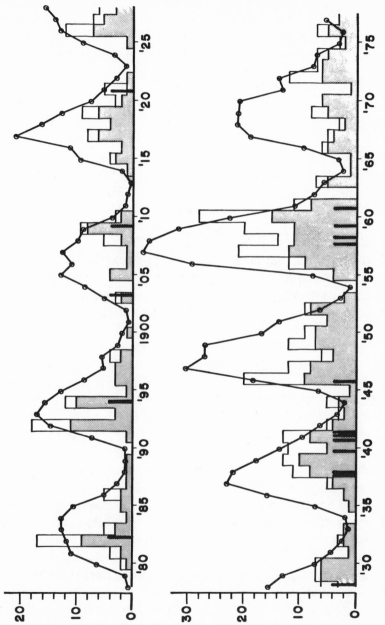

The sunspot cycle and major magnetic storms on earth during the past century. Open bars indicate the number of half-day periods each year when major geomagnetic storms were in progress. Shaded bars give the number of separate major geomagnetic storms each year. The heavy vertical lines indicate times of the seventeen "super" geomagnetic storms. Most of these super storms have occurred during the declining phase of a sunspot cycle. During the first seven years of a solar cycle, geomagnetic storms are usually associated with active regions (flares) and other short-lived phenomena on the sun; during the final four years, they are usually associated with coronal holes. (H. H. Sargent III/NOAA Space Environment Laboratory)

NUMBER OF PERIODS WITH aa-INDEX OF 100 OR MORE

instrumentally measured geomagnetic intensity where a "major" storm rates 100, the 1972 storm reached 220. Over the past century there have been seventeen geomagnetic storms that registered more than 350. Howard H. Sargent III, a solar scientist at the Boulder center, refers to these as "superstorms."

Such superstorms have occurred, on the average, once or twice each solar cycle. The 1972 storms happened during the declining phase of solar cycle 20. (Scientists begin the numbering with cycle 1 starting about the year 1755.) But the preceding cycle 19, which had the highest number of sunspots (another measure of solar activity) of any in the past one hundred years, produced four superstorms. All were during the cycle's declining phase in the late 1950s and early 1960s. And solar cycle 17 produced no fewer than six! They too happened when the cycle, as measured by sunspot activity, was declining during the late 1930s and early 1940s.

Auroral observations seem to confirm this view of abundant superstorms in the recent past. One somewhat subjective way to rate the magnitude of a magnetic storm on earth is to note the maximum southward extent* from which auroras were seen at the time. During the 1972 storm, there was at least one report of the aurora's being seen in Kentucky. But the atmospheric physicist Sidney Chapman once compiled a record of low-latitude auroras. He noted that auroras had been sighted in this century in Florida, Cuba, Hawaii, and other tropical locations. Most of these low-latitude auroras are also identified with superstorms. He even recorded that an aurora had been seen as far south as Singapore, 1° north of the equator! That was on September 25, 1909, the day of the greatest geomagnetic superstorm of the century. On the scale defining superstorms as those with a rating higher than 350, that storm reached 546!

What does all this mean for the future? It seems that we can expect to have one or two potentially highly disruptive superstorms each solar cycle, or roughly every eleven years. The record of the past century clearly shows that the most energetic solar flares and thus the strongest geomagnetic storms happen during the declining phase of the cycle, several years after the sunspots begin to recede from their peak numbers. For solar cycle 21, a very strong sunspot cycle that

---

*In the northern hemisphere. We tend, regrettably, to be northern hemisphere chauvinists. For the southern hemisphere the significant observation would be the maximum northward extent.

peaked in early 1980, that means the dangerous time for a severe flare-caused magnetic storm on earth would be the years 1982, 1983, and 1984. If solar cycle 21 were to produce a great geomagnetic storm at the same point in the cycle where the August 1972 storm occurred in cycle 20, the expected date would be (by adding 11.2 years) October 1983. The analogous time for cycle 22 would be December 1994.

But there's another factor too. The records seem to indicate that alternate solar cycles produce the most intense geomagnetic storms. Solar cycle 17 triggered six superstorms; cycle 18, only one. Solar cycle 19 produced four; cycle 20, during the early 1970s, none. The record for alternation doesn't look quite so good farther back in the century. But there are still good grounds for expecting that alternate cycles—those that happen to be odd-numbered—are more likely initiators of superstorms. And that gives further credence to the view that cycle 21 could still produce some major fireworks affecting earth before it reaches its minimum in 1986.

There is still another point to consider. Think back to the kinds of effects the great August 1972 storms had on society. They disrupted radio communications, caused voltage fluctuations in power and telephone systems, induced currents in undersea telephone cables, tripped circuit breakers, damaged telecommunications equipment, and caused a huge transformer to explode. All are essential components of our modern, complex technological civilization. I've seen no reports of significant computer malfunctions due to the storms, but effects on computers are another concern. There is even a report, not confirmed by our military, that solar-induced currents in the August 1972 storm triggered mines placed earlier by the U.S. Navy in a North Vietnamese harbor. The U. S. Air Force is concerned about the possibility of solar-induced interference not only with vital communications but with over-the-horizon radar and the effective operation of spy satellites. In fact, the Air Force jointly sponsors the Space Environment Services Center in Boulder, and receives all its reports by hot line.

We are talking about important effects on essential technological systems. They are systems that fail only at considerable threat of disruption or peril to today's society. We've grown to depend on the uninterrupted operation of these systems to an unprecedented degree in recent decades. In this sense our vulnerability to serious solar-caused geomagnetic disturbances is far greater now than ever before.

The aurora is now known to be a visible manifestation on earth of violent events on the sun. But it wasn't always that way. Medieval Europeans considered auroras omens of disaster. No wonder. They appear in the sky as a shimmery apparition, and at mid-latitudes the northern lights are dark red, a color naturally associated with blood and battle.

A contemporary description of an aurora widely seen over central Europe on January 12, 1570, referred to it as a portent and a "shocking prodigy." It left the people of Kuttenberg, Bohemia, fear-struck. There were "many burning torches like tapers and among these stood two great pillars, one towards the east and the other due north, so that the town appeared illuminated as if it were ablaze, the fire running down the two pillars from the clouds like drops of blood. And in order that this miraculous sign from God might be seen by the people, the night-watchmen on the towers sounded the alarm bells; and when the people saw it they were horrified and said that no such gruesome spectacle had been seen or heard of within living memory."

On occasion the glow is so brilliant towns are thought liter-ally to be ablaze. Seneca, whose descriptions of auroras in the early decades of the Christian era are free of any overtones of superstition, tells of one display seen from Rome. The blood-red glow in the west led everyone to believe the seaport town of Ostia at the mouth of the Tiber was on fire. The Emperor Tiberius dispatched his men to the town to help put out the flames.

One of the most spectacular auroras of this century occurred on the night of January 25, 1938. The sky was by good fortune clear that night. The BBC reported the events beginning on the evening radio newscasts, and much of Britain saw the display. Large expanses of the sky were blood-red, and ever-changing streamers of pale green, orange, and crimson appeared and disappeared again throughout the night.

An intense geomagnetic storm on the night of February 11, 1958, shifted the location of the aurora so far south that the auroral curtain was at times actually south of the U.S.–Canadian border. This left the polar sky temporarily devoid of the lights but gave much of the United States a spectacular rich red auroral sky. It was even seen as far south as Mexico City.

In today's satellite age, we now have views of the aurora never before available. Some of the more remarkable satellite photos I've seen show the North American continent at night, each city of any size clearly identifiable by its lights. And stretching across almost the entire width of Canada is the white glow of the aurora.

The story of the connection between auroras and the sun—or, for that matter, between any events on earth and violent activity of the sun—begins at 11:20 A.M. on September 1, 1859. That morning a thirty-three-year-old English solar astronomer named Richard C. Carrington, who had his own observatory in Surrey, was engaged in his daily task of mapping sunspots. He saw something happening on the sun.

"Two patches of intensely bright and white light broke out. . . . I thereupon noted down the time by the chronometer, and seeing the outburst to be very rapidly on the increase, and being somewhat flurried by the surprise, I hastily ran to call someone to witness the exhibition with me, and on returning within 60 seconds was mortified to find that it was already much changed and enfeebled. Very shortly afterwards the last trace was gone, and although I maintained a strict watch for nearly an hour no recurrence took place."

What Carrington had seen was the first eruption of a solar flare ever witnessed from earth. The detection of flares optically usually requires use of special filters on the telescope. These filters screen out all the sun's light except that of a certain wavelength—called the light of hydrogen alpha—that flares emit particularly strongly. Carrington's flare was rare. It emitted enough energy across the visible, or white light, portion of the spectrum to be seen by a human observer above the white glare of the sun's disk.

Four hours after midnight a great magnetic storm began. Auroras were seen over a large part of Europe. It was subsequently learned that the aurora was seen even in Honolulu that night and that there was extensive auroral activity in the southern hemisphere as well. Pioneers traveling in covered wagons westward across Kansas saw and recorded dramatic auroras that night. All over the world, including Australia, telegraph systems experienced problems that day. Currents were induced in telegraph lines. They were so strong that for two hours it was possible to send messages between Boston and Portland without any batteries, using only the induced current.

That November at the meeting of the Royal Astronomical Society Carrington reported his observation and hesitantly noted the coincidence of the subsequent magnetic storm. But there seemed no legitimate reason to connect the two events, and he cautioned against doing so. Said Carrington: "One swallow does not make a summer."

Nevertheless, in the latter half of the nineteenth century a similarity between the sunspot cycle and the auroral cycle became well accepted. Many astronomers as a result accepted a relationship between the solar cycle and magnetic disturbances on earth. But how such a connection could come about wasn't yet clear. It wasn't until well into the twentieth century that the majority of scientists became firmly convinced of the solar-terrestrial relationship.

As a result of efforts of many scientists over the past half century and especially during the past fifteen years, our understanding of the aurora and its connection to solar activity has improved dramatically. We now know that the aurora is a giant electrical discharge in the sky, operating much like a neon sign or like the cathode-ray tube in your TV set. And we now know that it is powered by a natural generator far above the earth whose output during an auroral discharge can total 1 trillion watts with a voltage of about 100,000 volts. The annual output is about 9 trillion kilowatt-hours, or about nine times the total annual consumption of electricity in the United States.

The generator responsible for this is called the solar wind-magnetosphere generator. Since the solar wind consists of a stream of charged particles from the sun, it is a conductor. Where the solar

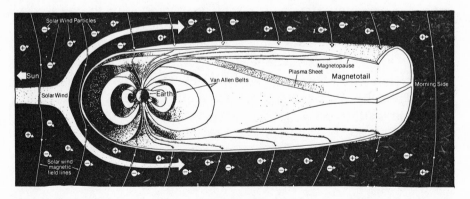

The solar wind-magnetosphere generator far above the earth produces prodigious quantities of electric current that leads to auroral displays. (S.-I. Akasofu/Alaska Geographic)

wind comes into contact with the influence of earth's magnetic field a large, elongate cavity is formed, posing an obstacle that the solar wind has to flow around like a ship producing a bow wave in the water. The cylindrical surface of this cavity is called the magnetosphere. Its surface is the generator.

The solar wind also carries lines of the sun's magnetic field with it, stretching them far out across interplanetary space to earth. In the last fifteen years scientists have learned that some of these lines of the sun's magnetic field connect up with lines of the earth's magnetic field across the surface of the magnetosphere. Now we know that when a conductor moves across a magnetic field, electricity is generated. The charged particles in the solar wind are the conductor. When they move across the connected magnetic-field lines at the boundary surface of the magnetosphere, electricity is generated on a large scale.

But the electricity has to get down into the earth's upper atmosphere to produce an aurora. That function is served by the bundles of earth's magnetic field lines that rise up from the polar regions to connect with the solar wind's magnetic-field lines. Under the special plasma conditions in which the gases of the upper atmosphere exist, the outer surface of this bundle of magnetic lines serves as "wires" for the transmission of current.

When the energetic electrons in this current collide with certain atoms and molecules in the upper atmosphere, the atoms and molecules emit their own characteristic light. When an oxygen atom is hit and excited to a higher energy state, it emits a greenish-white light. When a nitrogen molecule is hit, it emits a crimson light. The glow resulting from these emissions is the aurora. The polar lights are a visible manifestation of a series of events that begin on the sun and culminate in a dazzling display of color sixty to several hundred miles above the earth over the polar regions.

Auroral scientist S.-I. Akasofu of the University of Alaska notes that the bundle of magnetic field lines from the polar cap expands outward, like a funnel, as it extends up into space. And only the outer surface of this bundle carries currrent. This explains why the aurora is always observed to be a thin curtain of light and why this curtain always exists as a ring around the polar region, the auroral oval.

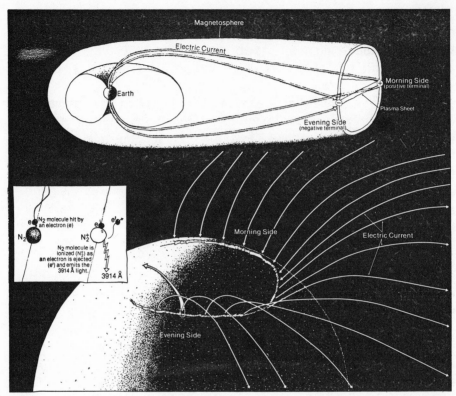

The auroral electric circuit. Earth's magnetic field lines conduct current generated by the solar wind-magnetosphere generator. Energetic electrons in this current ionize or excite molecules in the upper atmosphere, which emit colorful light along the auroral "curtain" surrounding the polar regions. (S.-I. Akasofu/Alaska Geographic)

The aurora shimmers or moves as the electric and magnetic fields in the magnetosphere change. And those fluctuations result from changes in the efficiency of the solar wind-magnetosphere generator. That efficiency is affected to a large degree by the particular direction of the solar wind magnetic field. But the efficiency is also proportional to the speed of the solar wind and to the square of the intensity of the solar wind magnetic field. In other words, a small boost in the intensity of the solar wind magnetic field produces a large increase in the generator's efficiency. When the generator efficiency is higher, the voltages and currents are higher and the resulting auroras are brighter. Also, when the solar wind and resulting geomagnetic storm are especially intense, the earth's magnetic field becomes distorted. The auroral oval then expands much farther than usual down into the lower latitudes, where it becomes a thrilling and awesome sight to millions of people who only rarely have an opportunity to see nature's colorful light show.

Flares aren't the whole story of sun-caused disturbances of the earth, however. Not by a long way. Frequent as solar flares are during the active and declining phases of the solar cycle, especially when all the minor ones are considered, they don't account for all the magnetic disturbances of earth observed. Exceptional activity, yes. More moderate, recurrent disturbances, no.

For the beginning of the story we go back more than half a century. In the 1920s radio (then called wireless) was just coming into wide use. By 1927, partly due to effects of magnetic phenomena on wireless transmissions, a recurrent twenty-seven-day interval in magnetic disturbance had been noticed. These moderate-sized recurrent disturbances were separate from the sporadic, large-scale disturbances associated with flares.

These recurrent storms tended to appear when the sun was free of any optical sign of activity. They recurred at the noticed interval whether the sun was active or "quiet." The twenty-seven-day period seemed to point to an association with the rotation of the sun. The sun's period of rotation as seen from earth is twenty-seven days.

In 1932, a German geophysicist, Julius Bartels, proposed the name M-regions (for "magnetically active" and "mysterious") for this still-unknown solar source of recurrent storms. This was the way things stood until the satellite era. The recurrent storms were shown to be associated with high-speed solar wind streams, but the source of these streams on the sun still was not known.

In the 1970s the mystery was solved. Many different lines of inquiry contributed to the solution. On March 7, 1970, there was an eclipse of the sun. The sun has a tenuous outer atmosphere called the corona that extends millions of miles out into space. Because the sun's surface is so bright, the corona becomes visible only when that light is cut off during a total solar eclipse. Photographs taken during the 1970 eclipse showed a dark cleft in the corona, as though a big chunk of it were missing.

Studies reported in 1973 of X-ray images of the sun made from rockets over a five-year period showed an abundance of holes in the corona. They were marked by low densities and weak magnetic fields. The magnetic fields converged outward from the sun as "open" rather than "closed" lines of force. The coronal holes seemed to be a kind of opening in the sun's magnetic field.

Theoretical studies had by now shown that just such an opening might allow high-speed solar wind particles to stream outward from the sun. But still no such link had been demonstrated.

Finally in 1973, the American scientist Allen S. Krieger and his colleagues managed to trace a stream of observed high-velocity particles back to its point of origin on the sun. That point was inside an observed coronal hole. There was little doubt. The high-speed wind stream had originated from the coronal hole. The discovery was reconfirmed by a number of observers in subsequent years. In 1976, for instance, scientists reported dramatic correlations between passages of coronal holes and magnetic disturbances on earth in the previous three years.

So, more than four decades after Bartels gave it the label M-region, the source on the sun of the twenty-seven-day recurrent geomagnetic disturbances was finally identified—the coronal holes. If flares are where energy and matter burst loose from the bounds of the sun in violent paroxysm, coronal holes are an open gate where they escape with barely an effort.

The Skylab manned missions in 1973 and 1974 were just in time to add measurably to our understanding of the sun, especially the coronal holes. The astronauts on board Skylab during all three manned phases of the mission took images of the sun every day in the hydrogen-alpha and ultraviolet portions of the spectrum. The large X-ray telescope on board Skylab took X-ray data of the sun regularly. Even in the periods when one crew had returned to earth from Skylab and the next crew had not yet been launched up to the lab, the X-ray photos continued to be taken automatically.*

The atlases of coronal holes compiled from these images

---

*There are two ironies about Skylab I think you might find interesting in this respect. The first is that it was one of the most successful and scientifically productive space missions ever conducted, yet it is remembered now in most people's minds more for the overdrawn concerns over where pieces of the huge spacecraft might fall when it returned to earth's atmosphere and disintegrated in the summer of 1979. The circuslike atmosphere and planetwide near-hysteria of that was really a little silly. The second irony was that Skylab, a laboratory in space that contributed so much to our understanding of the sun, was itself prematurely doomed by high activity on the sun. The reason its orbit deteriorated before we could take steps to save her was that unexpectedly high activity on the sun heated the top of our atmosphere, causing it to expand and exert more drag than usual on the huge abandoned spacecraft.

showed that they are an extremely important feature of the sun. Fully one fifth of the area of the sun was covered by coronal holes during the period of Skylab, a time of declining solar activity. Most of the holes were at high latitudes, both north and south, and were dubbed polar holes. Coronal holes were also found to be long-lasting features. Some of them lasted ten solar rotations (270 days) or more. They are thus among the longest-lived solar phenomena.

It is now well established that the high-speed streams of solar wind particles spiral out across space from coronal holes in the rotating sun like water from a revolving sprinkler. When the streams are aimed and timed so that they strike the earth, they can cause moderate disturbances of its magnetic field. As the sun rotates, the stream swings around to earth again every twenty-seven days, and we experience the moderate geomagnetic storms so long ago noted to recur at twenty-seven-day intervals.

Do flares and coronal holes exhaust the sources on the sun of magnetic convulsions on earth? Not quite. A remarkable but little-heralded contribution of Skylab was its confirmation of an extensive new class of solar activity called bright spots. Detectable in the X-ray and ultraviolet portions of the spectrum, these bright spots glitter all over the sun, like Christmas lights. They are smaller than sunspots, but they are also associated with small intense magnetic fields.

These bright spots blink on for a lifetime of about eight hours. Hundreds of them are scattered fairly evenly over the sun. About one in ten brightens explosively, like a miniature flare, for about ten minutes sometime during its brief life. The bright spots become more abundant as the numbers of sunspots and visibly active regions wane.

Bright spots may turn out to be an even more fundamental aspect of solar activity than sunspots. They appear to contain at least as much magnetic energy overall as sunspots and flare-producing active regions. Scientists now presume that it is these bright regions, detectable only from above the screen of our own atmosphere, that account for much of the variation in the solar wind and thus in the magnetic fluctuations at earth.

So we now have flares, coronal holes, and bright spots, all contributing to magnetic convulsions on earth. But even they don't account for all the disturbances observed. It turns out that there is still a fourth source. Huge bubble-like clouds of tenuous gas occasionally erupt from the sun's corona. These ejections of material and magnetic fields in the form of bubbles were another discovery of Skylab, which recorded more than a hundred of them, one as large as the disk of the sun. But the full significance of the process for earth didn't become confirmed until 1980.

Research scientists at the solar forecast facility in Boulder discovered in the late spring of that year that the two largest geomagnetic storms of solar cycle 21, then in the early part of its declining phase, appeared to be associated with these bubble ejections, called coronal transients. Although they are powerful particle eruptions, they lack the visual spectacle of solar flares—at least when seen from earth's surface. They are often marked simply by the sudden disappearance of dark bands of filaments of gas supported in the sun's corona by strong magnetic fields.

About 7 percent of all significant flares and 25 percent of all major flares had actually produced geomagnetic storm activity at that point in the sunspot cycle. But nearly 25 percent of all disappearing filaments were also associated with storms. These bubble ejections seem to be a strong rival to solar flares and coronal holes in the production of magnetic storms on earth.

Large flares and coronal transients are more frequent and violent after the peak of the solar cycle as measured by sunspot number. Bright spots are more numerous then too. And the coronal holes are largest and most abundant in the years just before the sunspot minimum. So, somewhat surprisingly, all four sources seem to be most active during the solar cycle's declining phase. This is another strong indication that the comings and goings of sunspots are at best only a superficial indicator of when to expect trouble from the sun.

Earth, we have seen, is awash in a solar sea. The usually gentle waves of particles and fields that flow past our planet, like an ocean current around a lonely island, can at virtually any time erupt into a violent storm, one made visible only by the shimmering and awesome beauty of the aurora. But in their effects on our modern civilization's electrical lifelines and communications nerve systems, these storms from the sun are a matter of more than aesthetic concern. Unlike the storms of land and sea, their drama is mostly hidden and silent. Yet they can affect us all, and we need to anticipate as best we can any unforeseen changes in their future actions.

3. THE INCONSTANT SUN

The history of science reveals many instances in which an apparent discovery, thought improbable or unlikely, was ignored to death only to be reborn at some later time in the light of fresh views and new evidence. Something like that has happened with the conception of the solar cycle. For most of the twentieth century we assumed it had been regular at all times in the past. The fact that it had not been so had been pointed out in the late nineteenth century. But no one seemed to take notice.

A German solar astronomer named Gustav Friedrich Wilhelm Spörer had been studying the distribution of sunspots with latitude. He checked back into the historical records. He was surprised. For a seventy-year period ending about 1716, there had been an almost total absence of spots on the sun. It was a remarkable interruption in the solar cycle. Spörer published two papers in the late 1880s, one in Germany and one in France, calling attention to the interruption. Spörer was well known. He and Carrington had independently discovered that sunspots occurred closer and closer to the sun's equator as the solar cycle reached toward maximum. But little note was taken of his papers about the cycle's apparent interruption.

One astronomer who did notice was an Englishman, E. Walter Maunder. Maunder was the superintendent of the solar department at Greenwich Observatory, where he made daily photographs of the sun and tabulated sunspot features. Maunder briefly summarized Spörer's two papers for the Royal Astronomical Society in 1890, and in 1893 began making his own search of the historical records. He was surprised to find that Spörer had been quite right. Sunspots had indeed been virtually absent from the sun for a prolonged period from 1645 to 1715. During nearly half that time, from 1672 to 1704, not a single spot was seen on the northern hemisphere of the sun. In 1705, when two sunspot groups were seen on the sun at the same time, it was the first time in sixty years that this had happened. The total number of spots observed during the entire seventy-year period was less than what astronomers now see in a single average year.

Maunder pointed out that when the astronomer Cassini reported the discovery of a sunspot in 1671, the editor of the journal carrying the report noted that no other sunspots "have been seen these many years that we know of." The editor even described the last previous sunspot seen, eleven years earlier, for those who had forgotten what one looked like. Maunder found that astronomy textbooks of the late 1700s had straightforwardly mentioned the late-seventeenth-century absence of sunspots.

Maunder—as had others before him—was about that time coming to the conclusion that solar activity seemed to be intimately tied to magnetic storms on earth. He had noted that in the nineteen years from 1873 to 1892, "the three magnetic storms which stand out pre-eminently above all others during that interval ... were simultaneous with the greatest development" of "three great sunspot displays." So Maunder fully realized that aberrations in the sun's behavior could very well have important effects on earth.

In 1894 Maunder published an article entitled "A Prolonged Sunspot Minimum," detailing all this evidence. It gained little attention. Neither did a similar article with the same title in 1922, which pointed out that what had happened on the sun in the recent historical past could happen again in the near future. Maunder died in 1928 at the age of seventy-six without having seen these ideas taken seriously. A two-column scientific biography of him published in 1974 doesn't even mention this aspect of his work.

The mid-1970s brought a new assessment. Enter at this time John A. Eddy. Jack Eddy is a solar physicist with a strong penchant for history. He is a senior scientist on the staff of the High Altitude Observatory of the National Center for Atmospheric Research, located in NCAR's modern E. M. Pei-designed monasterylike building on a mesa top overlooking Boulder, Colorado. A soft-spoken, unassuming man, Eddy combines the qualities of probing curiosity and a critical, evaluative mind necessary to the good scientist. He more than any other person is responsible for our newly awakened conception of an inconstant sun.

Eddy had seen occasional references to Maunder's assertions. From what he knew of the sun he felt the claim of a prolonged sunspot minimum wasn't true. He set out to put the idea to rest once and for all. As he said, "I felt it was time that we clear up the case of the missing sunspots, which had hung too long like a skeleton in the closet of solar physics." The subject had all the attributes of a good detective mystery. This intellectual challenge also attracted Eddy.

All the original accounts Spörer and Maunder had had at their disposal were still in existence. Eddy tracked them down and examined their dusty pages. "I found to my surprise that they were exactly as Maunder had represented them." They all indicated that from A.D. 1645 to 1715 sunspots were indeed a rarity. Eddy termed the historical evidence "striking." The name he applied to this prolonged absence of sunspots was quite appropriate: the "Maunder minimum."

Any contention that astronomers of the time did not have the means to observe sunspots, or did not look for them, just did not stand up. Galileo and other later astronomers had produced beautifully detailed drawings of sunspots prior to the Maunder minimum. Telescopes were in common use and produced commercially. Most of the fine observational detail known of sunspots today had already been recorded. Astronomers in England, France, Germany, Italy, and the world over, many of them quite respected, were perpetually peering at the sun with their telescopes. "It seems clear that the astronomers of the Maunder minimum period had the instruments, the knowledge, and the ability to recognize the presence or absence of even small spots on the sun," noted Eddy. "And we could add that it doesn't take much of a telescope to see a sunspot."

There is no way of knowing for sure whether a continuous watch was kept on the sun, but there is every reason to believe it was observed regularly. In this respect, Eddy found it significant that new sunspots were reported in the scientific literature as "discoveries." The sighting of a new spot or spot group was cause for writing a scientific paper about it. "I don't think that is a situation we could tolerate today," he noted with some amusement. Such a practice "would produce an intolerable glut of manuscripts in the minimum

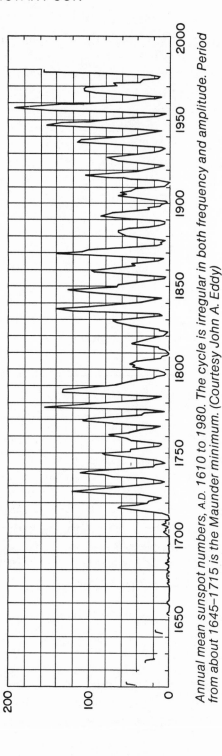

Annual mean sunspot numbers, A.D. 1610 to 1980. The cycle is irregular in both frequency and amplitude. Period from about 1645–1715 is the Maunder minimum. (Courtesy John A. Eddy)

years of the sunspot cycle and an avalanche in the years of maximum."

Still, as the astronomer Martin Rees has pointed out, "absence of evidence is not evidence of absence." Astronomers wanted more than a paucity of records of sunspots during the years in question before they could be certain the claimed minimum was not a result of incomplete or spurious data. It was this reservation that had kept Maunder's assertions on the shelf all this time.

Fortunately, a number of new tools for determining past sunspot activity were now available, some of them unknown in Maunder's time. They included better catalogs of historical auroras, compilations of Oriental sunspot observations, an understanding of tree-ring records, and the recent discovery that the proportions of atmospheric isotopes locked up in tree rings could be used as tracers of past solar activity. It was to these that Eddy turned.

Auroras, we have seen, are a product of solar activity. Any major increase in solar activity is inevitably followed by an increase in reported auroras. Prolonged periods without any aurora reports are a good indication that the sun is relatively inactive. In fact, the period from 1645 to 1715 was one of marked absence of auroras. Maunder himself had noted that. Eddy checked through catalogs of ancient auroras and confirmed that assessment. By today's standards, there should have been between 300 to 1,000 auroral nights in Europe between 1640 and 1700. But there are reports of only 77 auroras for the entire world from 1645 to 1715, and 20 of these came in a brief active period when sunspots were also seen. In one 37-year period of the Maunder minimum, not an aurora was reported anywhere. The few that were reported during the entire period were in far northern Europe. Not one was reported in London for a 63-year period ending in 1708.

The next London aurora was on March 15, 1716. It stimulated the astronomer Edmund Halley, of comet fame, to describe and explain it in a now-classic paper. Halley was then sixty years old, and he had never witnessed an aurora, even though he was a keen observer of the skies and had always wanted to see one.

Eddy concluded that the presence of the Maunder minimum in the auroral record was surely real.

Naked-eye sightings of sunspots in the Orient are another test of the reality of an extended sunspot minimum. The numbers are small, but they can be used as a very coarse indicator of past solar activity. A compilation of 143 sunspot sightings from Japan, Korea, and China, most from the third century A.D. until A.D. 1743, contained no sightings between 1639 and 1720. Eddy found the dates significant, "a Far East gap that matches Western Hemisphere data very well. No auroras were reported in the Far East in the same period either." Several auroras during that period have been found reported in Chinese civic records since Eddy's work, but the frequency is still very low.

There is still another visual clue to solar activity to check these correlations. During a solar eclipse the glare of the sun is eliminated and the sun's normally invisible tenuous outer atmosphere, the corona, becomes visible. The corona extends far out into space. You have seen the corona in photos of eclipses. A vast, irregular glowing area, it can be breathtakingly beautiful. We now know that the shape of the corona seen at eclipse varies with solar activity. When the sun has many spots the corona exhibits many long tapered streamers, a result of intense magnetic fields, that extend outward like the petals of a flower. When solar activity is low, the corona dims and contracts close to the sun's disk. Few streamers are seen.

Eddy went back over descriptions of the corona from eclipses of 1652, 1698, and 1706—all during the Maunder minimum. None describes the corona as showing structure. Not one mentions streamers. All indicate it was of very limited extent. No drawings were made. It is not until an eclipse in 1715, at the very end of the Maunder minimum, that distinct coronal structures are described emanating from the sun. It is only then that the corona is described in the modern form we are familiar with today. This is another strong indication something was awry with the sun prior to 1715.

But there is now another, even more profound indicator of past solar activity. It's almost like having a history book of the sun's past few thousand years. It comes from as unlikely a source as one might imagine—trees.

You've heard of archaeologists' use of carbon-14 dating of bones, charcoal, or wood. Carbon-14 is a radioactive version of normal carbon (which is carbon-12). It is produced in earth's upper atmosphere when cosmic rays from elsewhere in the universe strike atoms of nitrogen. So a certain small fraction of all the carbon atoms in the atmosphere are of this special kind called carbon-14. Trees absorb carbon in the form of carbon dioxide, in the same proportion of carbon-12 to carbon-14 that exists in the atmosphere while they are living. But when the trees die the carbon-14, being radioactive, decays at a known rate. By comparing the amount of it left in the wood of old trees with the amount presumably present at the time of its death, a good measure of the age of the tree is obtained. That is carbon-14 dating.

There is also tree-ring dating. Tree-ring dating depends on the fact that the width of the annual growth rings of a living tree varies from year to year according to whether climatic conditions are favorable or not. Series of years produce identifiable patterns, and by painstaking comparisons with trees of overlapping ages, experts have managed to compile a "pattern book" that allows them to determine the year a given ring was produced. In especially long lived trees this record extends back several thousand years.

But what has all this got to do with the sun? Well, it turns out that the amount of carbon-14 produced in the atmosphere varies from year to year as a result of solar activity. The solar wind streams out across the solar system and blocks a small portion of the cosmic rays from reaching earth. So when solar activity is high, slightly less carbon-14 is produced in the atmosphere. Trees growing at the time absorb less carbon-14. When solar activity is low, more carbon-14 is produced in the atmosphere. Trees growing then would have more of it to absorb. The ratio of carbon-14 to normal carbon in the annual growth rings of the tree can be measured in the laboratory. Since the age of the given ring is known by tree-ring dating, the carbon-14 content in that ring is thus a measure of whether solar activity was high, low, or normal at the time.

Time lags for the carbon-14 to descend from the upper atmosphere and be absorbed in the trees have to be considered. Other complications have to be taken into account too. But in general, it turns out that trees do contain a record of past solar activity. This tool became available just in time for Eddy to apply it to his study of the Maunder minimum. In fact, the Maunder minimum served as a kind of Rosetta stone that allowed the carbon-14 variations to be deciphered and interpreted for the first time as a record of solar activity farther back into the past.

If the sun had undergone a prolonged period of quiescence, it should show up as a time when the carbon-14 content of tree rings from that era was anomalously high. That is exactly what Eddy found.

The record showed a marked and prolonged increase in carbon-14 in trees that reached its maximum between about A.D. 1640 and 1720. The agreement with the time of the Maunder minimum was remarkable. This carbon-14 anomaly had been known for some years. It even had a name, the de Vries effect, after the Dutch scientist Hessel de Vries, who had first called geochemists' attention to it in 1958. It had been shown to be a worldwide effect. But until now, accounting for it had been a problem.

Eddy obtained data on relative deviation of carbon-14 from radiocarbon laboratories around the world. He plotted a curve of this variation, with increasing concentration downward, and on the same chart placed the records of sunspot numbers and of Oriental naked-eye sunspot observations. The correlation was obvious. Said Eddy: "The three quantities give a wholly consistent representation of the Maunder minimum."

The record also showed something else. There seemed to be not just one but three clear periods of solar anomaly in the past thousand years. The Maunder minimum, the most recent, was clear and apparent. But there was an earlier minimum, lasting from about A.D. 1460 to 1550. Here too sunspots recorded in the Orient were sparse and carbon-14 content in tree rings was high. Very few auroras were reported. This period of solar inactivity appeared to have reached its greatest depth in the early 1500s. Eddy named this period the Spörer minimum. At last the German astronomer's efforts to call attention to past erratic behavior of the sun had been duly recognized.

There also seemed to be a period of especially high solar activity in the twelfth and early thirteenth centuries A.D. There were unusual numbers of naked-eye sunspot reports from the Orient at this time, and the carbon-14 abundance was down. Eddy dubbed this the Grand Maximum. So not only was the Maunder minimum a real effect, it was only the most marked of several excursions in solar behavior over the past millennia. What did it all mean?

"We've finally shattered the principle of uniformitarianism for the sun," Eddy declared. "We've finally broken a block that has held us back. It was a comforting assumption that we took as fact."

He referred to the sun's irregular behavior as "a mark of a shaky machine." It appeared that we may have been self-centered and arrogant in assuming that the sun's present behavior is typical of its past and future.

"The existence of the Maunder minimum and the possibility of earlier fluctuations in solar behavior of similar magnitude imply

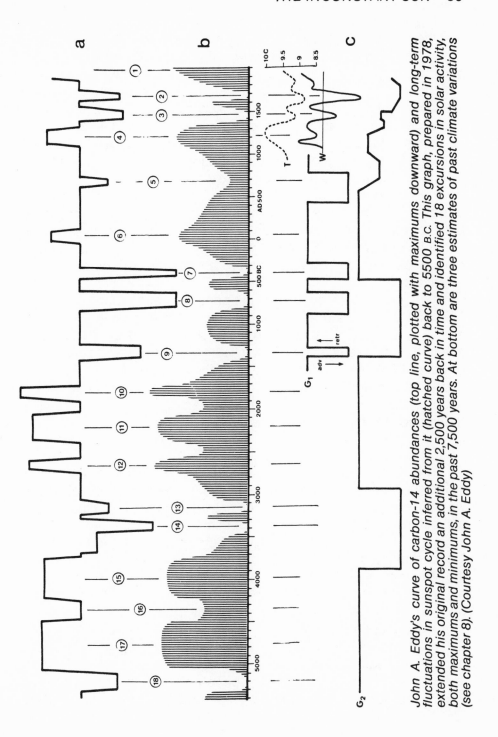

John A. Eddy's curve of carbon-14 abundances (top line, plotted with maximums downward) and long-term fluctuations in sunspot cycle inferred from it (hatched curve) back to 5500 B.C. This graph, prepared in 1978, extended his original record an additional 2,500 years back in time and identified 18 excursions in solar activity, both maximums and minimums, in the past 7,500 years. At bottom are three estimates of past climate variations (see chapter 8). (Courtesy John A. Eddy)

that the present cycle of solar activity may be unusual if not transitory," concluded Eddy. "For long periods in the historic past the pattern of solar behavior may have been completely different from our descriptions of the solar cycle today. There is good evidence that within the last millennium the sun has been both considerably less active and probably more active than we have seen it in the last 250 years. This opens the possibility of not only long-term changes in its radiative output, but also almost certain anomalies in the flow of atomic particles from the sun, with other inevitable terrestrial effects.

"The reality of the Maunder minimum and its implication of basic solar change may be but one more defeat in our long and losing battle of wanting to keep the sun perfect, and if not perfect, constant, and if inconstant, regular. Why the sun should be any of these when other stars are not is probably more a question for social than for physical science."

I was at the session of the annual meeting of the American Association for the Advancement of Science in Boston in late February 1976 when Eddy first reported these conclusions. The small meeting room was filled, and many listeners had to stand at the back or out in the hallway. When Eddy had finished, George B. Field, the session moderator and director of the Harvard-Smithsonian Center for Astrophysics, turned to the audience and said, "Maybe we've heard a turning point in the history of science."

Sunspot activity wasn't the only dramatic and visible change in the sun during the Maunder minimum. Incredible as it may seem, we have evidence that the sun's surface was rotating faster than usual just before the onset of the Maunder minimum.

Jack Eddy and colleagues Peter A. Gilman and Dorothy E. Trotter examined sunspot drawings made at two different times before the Maunder minimum to see if they could see any difference in rotation rates. Drawings made by Christoph Scheiner in 1625 and 1626 show a rotation pattern just like that of today. The sun's equator rotated at the same speed as today and the rotation rates at other latitudes differed by the same amounts as today. But drawings made by Johannes Hevelius from 1642 to 1645, just as the last sunspots were fading away to begin the Maunder period, show something quite different. The sun's equator had now speeded up. It was completing

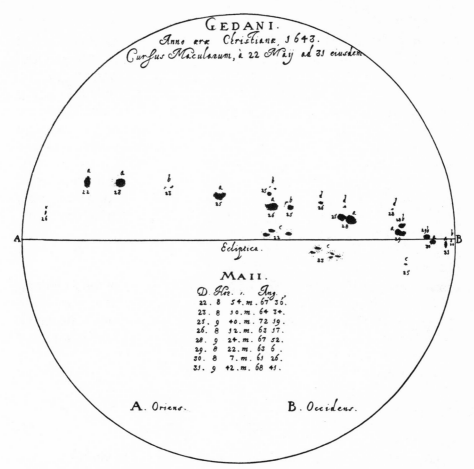

Drawing of sunspots in 1643 by Johannes Hevelius. (By permission of Houghton Library, Harvard University)

each rotation one full day faster than it had been just twenty years earlier. This also meant that the equator was rotating much faster than normally in comparison with other latitudes of the sun.

In 1978, astronomer Richard B. Herr of the University of Delaware analyzed even earlier sunspot drawings made during the earliest telescopically observed sunspot cycle. These drawings were made in England from 1611 to 1613 by Thomas Harriot. As we learned earlier, Scheiner and Harriot were two of the four co-discoverers of sunspots. Harriot's drawings are not especially precise, so their reliability is marginal. They seem to show an even slower solar rotation. If so, the acceleration up to the Maunder minimum was even more dramatic than that revealed between the observations of Scheiner and Hevelius.

What has solar rotation to do with an absence of solar activity? Which is cause and which is effect is not clear. But we do know that sunspots are areas of high magnetic intensity. And we know that their magnetic fields are produced from electrical currents generated through the action of a solar dynamo in which deep-seated magnetic fields interact with the differential rotation of matter in the outer layers of the sun. So extraordinary changes in the numbers of sunspots are likely to be associated with changes in magnetism or in the sun's differential rotation.

Eddy's initial reports on variable solar activity of the past went back to about A.D. 1000. But the record of past abundances of carbon-14 can be read in very old living trees, namely the bristlecone pine, the world's most long-lived tree, back almost to 3000 B.C. And it can be extended in preserved fragments of dead wood back to beyond 5000 B.C. With the help of specialists from radiocarbon and tree-ring laboratories, Eddy was soon able to trace the record of past activity of the sun back to the Bronze Age.

The results were striking. At least a dozen significant excursions in solar activity comparable to the Maunder minimum have occurred during the past five thousand years. Some are unusual increases in solar activity, some unusual decreases. Each lasts from fifty to several hundred years. The pattern is highly irregular.

To help in discussing all these past fluctuations in solar behavior, Eddy assigned names to each of the major solar excursions since 3000 B.C. based on the general period of human history in which the anomaly falls. We had the Sumerian maximum, for instance, at the time of the invention of writing, 4,500 years ago. There was a Stonehenge maximum, 4,300 years ago, an Egyptian minimum during the golden age of the New Kingdom, and a Roman maximum during the first century A.D.

Eddy now says he doubts that these names for the ancient solar excusions will be kept. "They were probably audacious anyway," he said. The carbon-14 records of the dates and durations of the solar events the names referred to are continually being refined and improved. The periods of the more recent episodes are also being refined, but their names are likely to stick To summarize, they are the

Modern maximum (roughly A.D. 1780 to the present), the Maunder minimum (A.D. 1640 to 1710), the Spörer minimum (A.D. 1400 to 1510), the Medieval maximum (A.D. 1120 to 1280), and the Medieval minimum (A.D. 640 to 710).

The ancient curves may explain some historical enigmas. We saw in the preceding chapter that during the times of the Romans there were some spectacular auroras. That agrees well with the radiocarbon record showing that during that period solar activity was high. But there was little mention of auroras by the Greeks of the fourth century B.C. even though theirs was an age of learning when early Greek interest in science and natural philosophy was developing. The radiocarbon record may show why. Solar activity was in a century-long minimum. Could it even be that Aristotle's conception of a perfect, unblemished sun arose in part because during his lifetime it apparently was indeed unmarked by sunspots?

The tree-ring radiocarbon record goes back even beyond the Bronze Age, although it becomes less reliable then. But it indicates marked maximums at about 6,000 and 6,500 years ago and a noticeable minimum about 7,200 years ago.

What does it all mean? Well, the record of the past 7,000 years shows clearly that the sun's behavior meanders noticeably, like a drunk staggering down the sidewalk. It's hard to detect any particular pattern in these long-term excursions. One thing for sure: the Maunder minimum is no unique anomaly. Such temporary shutdowns in solar activity seem to occur frequently. In fact, maybe they are the ordinary situation. Could it be that our present era of regularly recurring high solar activity every eleven years is the unusual state? Until the past year or so that seemed quite possible. We have had no direct evidence of an eleven-year cycle prior to the telescope. In August 1980, however, George L. Siscoe of the University of California at Los Angeles reported evidence of cyclic solar activity prior to the Maunder minimum based on six auroral catalogs that have been published within the last two decades. Siscoe found a roughly ten-year periodicity during aurorally rich intervals within the Middle Ages and the Renaissance. It looks much like the signal produced in modern times by the eleven-year solar activity cycle. It is interesting, and perhaps somewhat comforting, to know that the eleven-year cycle can be identified prior to the Maunder minimum and the invention of the telescope. Nevertheless, in the perspective of the longer-term irregular variations back to the Bronze Age revealed by carbon-14 fluctuations, the eleven-year cycle shrinks in significance. Perhaps we've been drastically misled by it. The newly identified gross changes in solar behavior lasting hundreds to thousands of years may be far more important. Jack Eddy said it well:

"The New Solar Physics tells us that the eleven-year cycle is but a ripple on an ocean of great and sweeping tides. It suggests we step back and look instead at the longer-term changes, when the sun drifts in and out of eras like the Maunder minimum. It says that these changes may be the more fundamental on the sun, the more indicative of changes in the sun's energetic, radiative output, and the more important in terrestrial effect."

In 1980, results of a new study of the carbon-14 record as an indicator of the solar cycle were published. Minze Stuiver, a respected radiocarbon expert at the University of Washington, and colleague Paul D. Quay had developed a way to measure the amount of carbon-14 in tree rings to a precision never before possible. Their samples were taken from Douglas fir trees from five locations in the Pacific Northwest. These trees had quite wide annual rings, which is important in providing enough material from each ring to allow the higher-precision measurement.

It was important to see if the previous radiocarbon evidence for past solar variations would stand up to this new and more stringent test. It did. Stuiver and Quay first calibrated their carbon-14 findings against the modern historical record of sunspots and found good agreement. When solar activity was high, carbon-14 levels were low, just as expected. They found even better agreement between the carbon-14 levels in tree rings and modern records of geomagnetic variations on earth, another good sign that they were measuring a real solar effect. They then used their new carbon-14 data to produce a new profile of solar activity over the past 850 years. It confirmed the timing and magnitude in the ups and downs of solar activity Eddy had identified. The Maunder minimum is there, and the Spörer minimum too. Stuiver even found clear signs of a previously unidentified minimum, lasting from about A.D. 1282 to 1342. Naked-eye sunspot records show an absence of sunspots during that time too. And it shows up in Siscoe's compilations of auroral records. Stuiver and Quay named this new period the Wolf minimum, after the nineteenth-century Swiss astronomer Rudolf Wolf, who originated the system of counting sunspots still in use today. The Stuiver study found that solar activity was extremely low for about 60 years during the Wolf minimum, 118 years

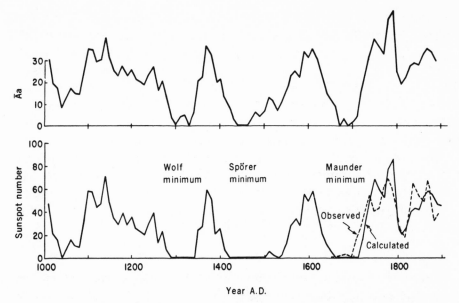

The most finely detailed record yet attained of sunspot numbers and geomagnetic activity over the past millennium. Both curves were calculated from
the carbon-14 production record in tree rings. (Minze Stuiver and Paul D.
Quay)

during the Spörer minimum, and about 60 years during the Maunder
minimum.

Thus for more than half the 432 years from the time of Dante
and Marco Polo to the boyhood of Benjamin Franklin and the bur-
geoning development of the American Colonies, the sun was nearly
devoid of sunspots and in general acting quite differently from the way
it behaved during the most recent 250 years.

Even these past 250 years haven't had the sun quite as
consistent as we used to think. The solar cycles have occurred regularly during this modern period, but their strength has varied. The
charts of sunspot cycles, for instance, show that the peaks of the 11-
year cycles were quite lower than usual during the early 1800s and
around 1900. This led to the idea that there is some variation in
sunspot activity that recurs at 80- or 100-year intervals. It's sometimes
called the Gleissberg cycle, after the German astronomer who wrote of
it in the 1940s.

Whether it's a cycle or not, two American solar scientists,
Joan Feynman and S. M. Silverman, examined a very reliable set of
records of auroral sightings in Sweden. They reported in 1980 that

there was indeed a sparsity of auroras from 1809 to 1815, and the few that were seen were in the extreme northern latitudes. They conclude that the solar wind was especially weak in some sense during this period. The record seems to indicate that the wind was strong from 1771 to 1793, became weaker with the pronounced minimum from 1809 to 1815, and then returned to a strong state from 1837 to 1858. It was weak again in 1901. They conclude that the strength of the sunspot cycle and the solar wind must both be related to some underlying 80- to 100-year variability of the sun. They feel this longer Gleissberg variation in strength of the solar wind may, in fact, be just as fundamental as the 11-year sunspot cycle or 22-year magnetic cycle of the sun. Siscoe's study also found clear signs of a roughly 80-year cycle in solar variability during the Middle Ages.

The sun is the only star in our solar system. (It would be quite a strange and different solar system if it weren't!) That's perhaps fortunate for the development of life here, but it does limit our understanding of these matters a bit. We have no other star equally handy for comparison. It's hard to know whether these irregularities in the sun's activity are something normal in the life of a star.

Well, no other star is in the same vicinity, but there are billions of stars out there. It turns out that even though they are at great distances, we can still tell something about their activity by analyzing their emanations.

In the past few years orbiting space observatories, especially the one known as Einstein,* have shown that an unexpectedly wide variety of stars are active at levels at least as great as those of the sun, and in some cases far greater.

The satellite observations show many signs of magnetically active features on stars. Some sunlike stars, for instance, apparently have large fractions of their surfaces covered with spots (starspots?). Flares explode away from the surface, just as on the sun, and their outer atmospheres emit X rays, often at intensities much greater than those we see on the sun. In contrast, other sunlike stars are much weaker than the sun in surface activity and in X-ray production from their atmosphere.

---

*It was launched in late 1978, as we were preparing to celebrate the centennial of Albert Einstein's birth.

A star known as BY Draconis exhibits semiregular brightness variations that some scientists now think are due to the rotation of sizable groups of starspots onto the part of its disk facing us. If the scientists are right, BY Draconis would have to have 30 percent of its surface covered with spots at these times. By comparison sunspots are quite sparse: only about 1 percent of the solar disk is ever covered by sunspots. BY Draconis and other stars like it also exhibit frequent flaring activity.

Scientists at the Harvard-Smithsonian Center for Astrophysics have examined Harvard's eighty-year collection of photographic plates of stars like BY Draconis. They see modulations in the stars' brightness with periods up to decades. They think these long-term variations may be due to starspot cycles, analogous to the sun's eleven-year sunspot cycle.

One such star, HD 224085, varies in brightness over periods of several years. But between the years 1900 and 1940, its brightness remained constant. Some of the scientists think this is a sign the star underwent a prolonged starspot minimum, analogous to our sun's Maunder minimum.

One astronomer, Olin C. Wilson of the Hale Observatories in California, has used the 100-inch telescope on Mount Wilson to detect stellar activity cycles on other stars. He observes these stars using a filter that senses only two certain wavelengths emitted by calcium atoms. It is known from studies of the sun that calcium emissions are produced in magnetically active regions and that they vary in step with the sunspot cycle. So on other stars the calcium-light variations should indicate magnetic and starspot cycles.

Wilson sees cyclic variations in a number of stars. And some other kinds of variation may prove to be cyclic if observed long enough. In addition, almost all the stars he observes show rapid variations—changes from one month to the next or even between successive days. He hasn't yet been able to determine whether these rapid variations are regular or not.

Wilson and his collaborators, Arthur Vaughn and Dimitri Mihalis, consider the evidence for the longer-term variations to be quite good. They see cyclical periods ranging from about seven years to probably at least twice as long. The sun's eleven-year cycle thus falls well within this "normal" range. And they consider these observed cycles to be quite analogous to those of the sun.

So far this research hasn't been able to cast any light on

whether the stars suffer interruptions in their cycles as the sun has. But it does indicate that perhaps half or more sun-type stars may show cyclic behavior. The observations are continuing, and they should eventually be able to answer many exciting questions.

Scientists involved in the program of observations now in progress hope to be able to identify individual active regions rotating with stars. From that they should be able to identify the rotation periods and the degree of differential rotation, which is what drives the dynamo that produces magnetic activity and starspot groups. Also, by examining cycles on stars like the sun but of different ages, they should be able to learn how the cycles vary as a star gets older. This should tell us something about what to expect from the sun in the future. And when enough stars have been observed for long enough times, the statistics should show how typical are the interruptions we've frequently seen on the sun, from the Maunder minimum on back.

The past half-dozen years have shaken our conception of the sun. We've learned conclusively that the sun's cycles of activity weaken and reappear at intervals that appear to be irregular. It is a challenging realization, because of the intellectual mystery, and an unsettling one, because we know the sun can have profound effects on earth. But the sun has only started revealing its secrets. The study of distant stars like the sun may eventually place all this seemingly aberrant solar behavior in a better cosmic context. And that, in itself, should be of some comfort.

4. A PAUCITY OF GHOSTS

In the late 1920s, physicists exploring the inner nature of matter were puzzled. In the decay of radioactive atoms, the particles emitted didn't carry enough energy to account for all the mass lost by the radioactive nucleus. This imbalance of input and output seemingly violated the law of conservation of energy, one of the great principles of science. Experiments had always shown that nature keeps its energy books strictly in balance.

What was wrong? Niels Bohr, the great Danish physicist, felt that the conservation laws must simply be invalid for this kind of reaction, a highly troubling conclusion to say the least. Then along came Wolfgang Pauli.

Pauli was a young Austrian-born physicist. Something of a youthful prodigy in science, Pauli had written highly praised articles on the then-new theory of relativity while only a teenager and a brilliant book on the subject when only twenty-one. In 1925, when only twenty-five, he announced what is still known as the Pauli exclusion principle, a conceptual breakthrough in understanding the arrangement of electrons in the atom that helped explain the structures and properties of atoms from hydrogen to uranium. He would win the Nobel prize in physics twenty years later for that.

Pauli was an affable and amusing man. He brought both

laughter and new ideas everywhere he went. By all accounts, he was also quite clumsy, a peril to any laboratory equipment unfortunate enough to be in his proximity. Colleagues jested that their lab equipment broke even as he walked in through the doorway, a phenomenon physicist George Gamow jokingly labeled the Pauli Effect, "a mysterious phenomenon which is not, and probably never will, be understood on a purely materialistic basis." Fortunately, Pauli was a theoretical, not an experimental, physicist, and none of that mattered.

In 1928, Pauli became a professor at the University of Zurich. He turned his attention to the radioactive-decay problem. In 1931 he proposed that the mystery could be solved and the conservation law saved if some previously undetected particle, as well as the usual electrons, were emitted during the decay process. It would have to be a strange particle indeed. It could carry energy and have "spin." But it would have no electric charge and no mass (or none to speak of). Its existence, however, would balance the energy books.

Pauli informally called the particle the "neutron" because of its neutral charge, but that name was formally preempted in 1932 and applied to a quite massive neutral particle that had just been discovered in the nucleus of the atom. The Italian physicist Enrico Fermi then suggested that the hypothetical Pauli particle take its name from the diminutive form of the Italian word for the neutron, and thus it was named the "neutrino," meaning "little neutral one."

Did the neutrino really exist or was it just a physicists' bookkeeping trick to avoid the trauma of undermining the conservation laws? Physicists argued the question, but a variety of circumstantial evidence soon indicated that it must exist. Reactions involving the newly discovered neutron required the neutrino. When a proton is converted into a neutron, the products are a neutron, a particle called the positron, and a neutrino. Without our going into the details, that neutrino brings the book for that reaction into balance.

But what an elusive ghost of a particle it is! It has zero charge and zero mass, and it must always travel at the speed of light. It is virtually undetectable. It doesn't interact with matter the way most other particles do—electrically or gravitationally or by the force that holds the nucleus together. All these quirks make it the most penetrating subatomic particle known. Scientists have calculated that the average neutrino could pass through 100 light-years (600 trillion miles) of solid lead with only a 50 percent chance of being absorbed.

Neutrinos pass right through the solid earth as though it weren't there. Only about one in 10 billion neutrinos zipping through the earth reacts with any matter there.

It's hard enough to imagine such a particle. It's harder yet to detect it. Yet the neutrino (actually a form called the antineutrino, but the distinction is unimportant here) was finally detected in 1956. Two Los Alamos Scientific Laboratory scientists, Clyde L. Cowan and Frederick Reines, managed to capture it in an elaborate experiment in South Carolina using streams of neutrons from the Atomic Energy Commission's powerful reactor complex on the shores of the Savannah River. The breakdown of these neutrons should produce an antineutrino. If antineutrinos were produced in huge enough quantities, a very few might interact with the protons in the hydrogen atoms in a huge tank of water. The products of those reactions could be trapped by the experimenters' detection devices. The devices did monitor these products at just the energy levels calculated. The neutrino had finally been trapped.

What has all this to do with the sun? Several of the reactions by which the sun produces its energy should involve release of neutrinos. A complex experiment to detect these solar neutrinos operated throughout the 1970s, and it didn't find them in nearly the amounts expected. This led to a feeling of crisis among scientists. Something seemed to be awry with their theories about how the sun works. But let's backtrack just a bit.

The sun produces its energy by the process we call thermonuclear fusion, the conversion of hydrogen into helium. In a series of three steps, four protons (each the nucleus of a single hydrogen atom) become fused, under the enormous temperatures and pressures deep inside the sun, into two protons and two neutrons (a helium nucleus). In these transmutations a slight amount of mass is lost. It amounts to 0.7 percent of the mass of the original protons. It is the conversion of this mass to energy, via Einstein's famous equation $E = mc^2$, that powers the sun and, incidentally, warms and lights our planet, makes possible all plant growth through photosynthesis, produces rain, winds, and the weather, and otherwise makes it rather convenient for us to live here on earth.

The energy released this way is almost inconceivable. Einstein's equation shows that even a small amount of mass is converted to a huge amount of energy; the *c* in the equation refers to the speed of light, which is a large number in itself. The speed of light squared is far larger. So even though only a small fraction of the mass in the conversion reaction is converted to energy and only a small fraction of the total mass of the sun is involved at any time in the thermonuclear burning, the total amount of energy released is monumental. Each second the sun produces enough energy to supply the needs of the United States for electricity for 50 million years (or a little less, if we don't conserve). That is the energy of 100 billion one-megaton hydrogen bombs exploding each second.

Some 4.2 billion kilograms (4 million tons) of the sun's mass gets converted to energy this way each second. That sounds like a lot. But the sun has a mass of 2×10^{30} kilograms (2 followed by 30 zeroes). Even with 31,557,600 seconds in a year, that works out to less than 1 part in 10 trillion of the sun's mass being converted to energy this way in a year. It's not in imminent danger of using up all its mass as fuel.

There is no way to observe this process directly, of course. The solar nuclear furnace is deep within its core, and the cooler outer layers of the sun forever obscure its workings from our view. The matter surrounding the core is, in fact, quite opaque. No particles originally produced there remain long in their original form. A particle of light inside the sun can travel only about the width of your little finger before it is absorbed. It takes millions of years, and endless absorptions and reemissions, for the energy of that particle to reach the surface of the sun.

The neutrino, however, doesn't have that problem. Due to its strange properties it has little likelihood of interacting with any matter in the sun. It races unhindered at the speed of light outward from the solar interior, reaching the surface in a little over two seconds even from the very center of the sun, and races out into space. Eight minutes later it reaches the earth's distance from the sun.

Half of the protons involved in the fusion process decay into neutrons, and each such decay results in the emission of a neutrino. The number of protons in a kilogram of hydrogen is roughly the number 6 followed by 26 zeroes, so you can see we are talking about the production of neutrinos in copious quantities. Some 8×10^{28} of them are passing effortlessly through the earth each second. In fact, about 200 trillion are passing right through you this second, whether you're indoors or out, or it's day or night (for as we've seen the solid earth is no screen to solar neutrinos).

Neutrinos from the sun provide the only direct way to test our detailed understanding of how the sun generates its energy. Being the only particle that reaches us untransformed from the nuclear reactions in the core, they represent our one way to "look" into the sun's interior and see if those processes are working the way we think they are. They are our one messenger from the center of the sun.

But there's a problem. Neutrinos from the main proton-proton reaction in the sun are produced in plentiful quantities, true, but their energy is only about half that necessary for them to be detectable by any means yet available.

All is not lost, however. Our best models of the sun predict that there should be certain other fusion reactions going on in the sun. Although less common, they should produce more energetic neutrinos. About one time in 400, for instance, two protons and an electron combine and release a neutrino with about three times the energy of those produced in the main proton-proton reaction. That makes these neutrinos just barely detectable.

But there's another side reaction that's even better. It occurs about one time in 5,000. A helium nucleus (helium-4) fuses with a lighter version of helium (helium-3) to form an element called beryllium-7. The beryllium-7 combines with a proton to form boron-8. This boron then decays into a radioactive isotope, becoming beryllium-8, with the release of a positron and a particularly energetic neutrino. All that concerns us here is that neutrino. It is up to 33 times as energetic as the neutrinos produced in the sun's main reaction, and thus should theoretically be detectable.

Is it? How, after all, do you capture a fleet ghost? How do you detect a massless, chargeless particle that passes through the entire earth as though it weren't there? It would also pass through any detector. True. But even though almost all neutrinos pass effortlessly through matter, every rare once in a while one will nevertheless happen to hit the nucleus of an atom so squarely it will interact. Given the large predicted flow of neutrinos from the sun, a very few will interact with something. The trick is to be able to capture that rare event.

Well, neutrinos can be captured by the inverse of the decay process by which they are formed. Just as a neutrino is released when a proton converts to a neutron, the absorption of a neutrino will convert a neutron to a proton. If, for instance, the nucleus of chlorine-37 is struck squarely by a neutrino, one of the neutrons there will be converted to a proton. The resulting atom we call argon-37. It does not combine with other atoms into molecules and it is radioactive. Both attributes make its presence fairly easy to detect.

So chlorine might make a good neutrino detector. Lots of chlorine. Say, 100,000 gallons of chlorine. But it would not be enough to attach argon-37 detectors to a swimming pool full of chlorine. Cosmic rays (high-enegy particles produced by violent events elsewhere in the universe) can also produce argon-37 when they impinge on chlorine atoms. So the vast vat of chlorine would have to be placed either at the bottom of the ocean or deep underground, where cosmic rays can't penetrate. If underground, it would have to be surrounded by a "blanket" of water which would absorb radiation emitted by natural radioactive elements in rock.

If you fulfilled all these requirements and had all the devices necessary for filtering out and detecting argon-37 atoms, you would have a neutrino detector, or "neutrino telescope," as it's sometimes called. It's a way of finding out what's going on in the heart of the sun. And that's what Ray Davis designed and built.

Raymond Davis, Jr., is a nuclear chemist who has spent his entire professional career with the Brookhaven National Laboratory in New York. And a good part of that time he's spent seeking solar neutrinos. Not at Brookhaven, but nearly a mile beneath ground in a gold mine near Lead, South Dakota. The Homestake Gold Mining Company excavated a large cavity for Davis's neutrino observatory.

There, safe from the intrusion of all but the most powerful stray cosmic rays, rests a tank filled with 100,000 gallons, or 610 tons, of common cleaning fluid, perchloroethylene, a cheap and abundant source of chlorine. The tank is allowed to stand for 35 to 100 days at a time. Then a chemical extraction system collects any gas and an instrument examines it for presence of any atoms of radioactive argon-37, a sure sign a neutrino from the sun had entered. A neutrino could be expected to react with a chlorine atom only a little over once a day—somewhere between 1.1 and 1.3 times—so after a 100-day experimental run the tank should contain at least 110 atoms of argon-37—if astronomers' suppositions about the sun are correct. The equipment is capable not only of detecting the presence of that few atoms but of counting them.

The experiment was run frequently throughout the 1970s—more than three dozen times, in fact. The results were increasingly disturbing. Averaged over the whole period the number of argon-37 atoms produced came to only about 0.4 a day, or one every two and a half days. To simplify matters, scientists developed a unit of reaction rate called the solar neutrino unit, or SNU. Theory predicted a neutrino capture rate of about 6 SNU; only about 2 SNU was observed. In other words only about a third the expected number of solar neutrinos have been observed.

This has been a very careful experiment. Davis is respected as a cautious investigator. Time after time he has brought the work before his scientific colleagues and sought their advice and criticism. All conceivable flaws in the experimental design have been considered. Improvements have been made. The few cosmic rays that managed to reach the depth of the observatory are detected separately and their effects found insignificant. Throughout the mid- and late 1970s the result gained more and more solidity. Solar physicists have had to face it. The result is real: neutrinos from the sun have not been observed in nearly the expected numbers. What is wrong? Are astronomers' views about processes inside stars seriously in error?

All sorts of suggestions have been advanced, some only half seriously. There might be huge magnetic fields in the solar center. Hydrogen and helium might not be well mixed. Perhaps the interior composition of the sun is far different than we thought.

One astrophysicist even suggested that there is a black hole at the center of the sun, an object so unbelievably dense that half of the sun's luminosity is generated when the black hole accretes mass from the solar gas surrounding it.

It has even been proposed that the sun has stopped shining, or is in a transition stage between full and diminished energy generation. A shortage of neutrinos would be the first clue to such a happening, since their flow is a consequence of events only 8 minutes ago in the center of the sun while the visible light we see had its origins in events millions of years ago. Thus the solar furnace could have switched off anytime in the recent past and we still wouldn't know about it except for an absence or shortage of neutrinos.

Astrophysicists are an imaginative sort. They have to be to consider the wonderful and awesome phenomena of quasars, pulsars, black holes, and exploding supernovas they deal with. There's been no shortage of speculation about the solar neutrinos. But it is speculation in the absence of facts. Davis's theoretician colleague John N. Bahcall of the Institute for Advanced Study at Princeton has dubbed some of the suggestions "cocktail-party solutions." In other words, they were contrived solely to dispose of the missing-neutrino problem in the absence of any evidence to support them or to suggest even whether they're likely.

Throughout all this unsettling time, many physicists have felt that the astronomers don't understand the sun as well as they think they do. And many astronomers have decided that there must be something wrong with the basic physics.

And that brings us to one more possibility. Maybe our understanding of the neutrino has been wrong. Perhaps this pesky little particle we thought we knew so well has been throwing us a curve. Suppose something happens to it during its short, swift flight from the solar interior to Ray Davis's subsurface observatory. This was always recognized as a possibility, but there was no evidence anything of the sort happened. In considering all the speculations to explain their neutrino data several years ago, for instance, Bahcall and Davis noted the suggestion that the neutrino may behave differently in its flight of the 1.5 billion meters from sun to earth from the way it does in laboratory measurements over a distance of less than a meter. They mentioned the proposal that the neutrino may indeed have a tiny amount of mass, which might allow it to decay into some unknown particle. But, they added, correctly (at the time): "The latter suggestion has not been taken very seriously by most physicists because there is no independent evidence for the postulated decay product...."

In the spring of 1980, Frederick Reines announced that he may have found that evidence. He reported at the American Physical Society meeting in Washington that he and colleagues had found evidence that the neutrino changes its form in flight. To do so, it must have some mass after all. It was a stunning development, one that would have ramifications across the physical sciences. And it could explain why so few neutrinos have been observed from the sun.

To find the evidence, Reines had returned to the site of the experimental detection of the neutrino by Cowan and him in 1956, the Savannah River reactor in South Carolina. Since the discovery of neutrinos, scientists had now found that they came in three varieties, or "flavors." The ordinary neutrinos, those predicted by Pauli, are always produced in association with an electron. These electron neutrinos are the kind generated in the sun. But there are also neutrinos associated with two heavy relatives of the electron, the muon and the tau particle.

Reines, now at the University of California at Irvine, and Irvine colleagues Henry Sobel and Elaine Pasierb, placed a detector full of heavy water (deuterium) thirty-seven feet from the Savannah reactor core. They then monitored interactions between the deuterium nuclei and neutrinos spewing from the reactor. The results were puzzling. It took the Irvine physicists a while to understand what their data were apparently telling them. As Reines put it when he reported their conclusions to the Physical Society:

"Six weeks ago there occurred to us a powerful way to look at the data. If one analyzes the data in a special way, uncertainties melt away like snow on a warm day."

There were two kinds of reactions going on. One takes place only with electron neutrinos. The other can occur with any variety of neutrino. Fewer electron neutrinos than expected were observed. Less than half the expected number, in fact. Yet the *total* number of neutrinos observed was just as expected. The conclusion had to be faced. "This means that the electron antineutrinos appear to be changing into another type of neutrino," Reines reported.

The analysis seems to indicate that what is ostensibly an electron neutrino when it comes out of the reactor can change itself to a muon neutrino and back perhaps more than once on the way to the detector. The neutrino, in other words, seems to oscillate from one form to another during flight. It is as though, in an analogy Reines offered, a dog changed into a cat and back into a dog again as it walked down the street. Said Reines: "The universe is not the way we thought."

In science, no one experiment stands alone. Only when it has been thoroughly critiqued by others and when other investigators have been able to obtain the same results in independently conducted experiments does an experimental result begin to achieve the status of a "discovery." Efforts to search for evidence of neutrino oscillation were soon under way elsewhere. Some of the early signs seemed to support the Reines finding, others didn't. One physicist called the experiment a brute-force, heroic effort unfortunately lacking in any internal checks. If the findings of Reines and his colleagues *do* stand the test of time and further experiment, however, the implications would be enormous.

To change from one form to another, a neutrino must have mass. And if neutrinos have mass, even only a very small amount, their overwhelming numbers would mean they would outweigh everything else in the universe put together. And this would give the universe enough mass so that mutual gravitational attraction will eventually stop its present expansion and it will begin to contract. The findings would also further enliven the efforts of physicists to understand the basic structure of matter.

But most important for our purposes, the result might explain why Ray Davis hadn't been seeing all the solar neutrinos scientists expected. When they pass through his underground chlorine tank, many of them may temporarily be in the form of muon neutrinos, not electron neutrinos, and thus pass through undetected. If so, the mystery of the missing neutrinos would apparently be solved. In that case we could say the fault, dear Davis, is not in our sun, but in our neutrinos.

Yet how can we know for sure? If Frederick Reines's 1980 Savannah River experiment is confirmed by subsequent work, that, you might think, should settle the matter. The neutrinos themselves would be the culprit. But things are never that simple. "Plainly we are in for a period of lively argument," physicists A. De Rujula and Sheldon Glashow wrote in *Nature* in August 1980. "The interpretation of the experiments so far reported is both uncertain and controversial . . . the hope must be that people will now design the experiments that must and should be carried out to settle this intriguing issue." They called the Reines experiment and another, more preliminary one at the CERN laboratory in Geneva "teasers for the more precise experiments that can and must be done."

Such experiments were being designed. Massive equipment designed to test for neutrino oscillations in more refined ways was soon under construction in a warehouse near Irvine. It would eventually be transported to New Mexico (Los Alamos) and to South Carolina (the Savannah River Plant again), where the actual experiments would be conducted.

Even if neutrinos are conclusively shown to oscillate, the solar-neutrino mystery would not necessarily be solved. That was clearly evident in a paper by John Bahcall and eight colleagues from the Institute for Advanced Study, Los Alamos, Yale, Lawrence Livermore Laboratory, and UCLA written shortly after the Reines experiment was announced. Published in the September 15, 1980, *Physical*

Review Letters, it reported new calculations by Bahcall's group on the rate of solar neutrinos theoretically expected to be captured in the chlorine-37 experiment. It was an update of their last previous calculations published seven years earlier. In the intervening years, physicists had gained better knowledge of the opacity of the sun's interior to radiation (it was about 15 percent higher than previously thought), an improved value for the luminosity of the sun, and new measurements of the lifetime of the neutron, among other things. All these affected the calculations.

The result was that the theoretically expected rate of neutrino capture was now significantly higher than their estimate of 6 SNU published in 1973. It now was about 7 to 8 SNU, or somewhere between 2.6 and 4 times (depending upon uncertainties) the number of neutrinos actually observed in the chlorine-37 experiments.

This discrepancy, Bahcall and colleagues said, was now so large that neutrino oscillations, even if confirmed, probably could not account for all of it. "It is difficult to resolve the difference between predictions based on the solar models and observations solely by invoking neutrino oscillations, if there are only three kinds of neutrinos coupled to each other," they wrote. They pointed to estimates that neutrino oscillations could probably account for a factor-of-2 discrepancy between observations and theory. Their new best estimate for the actual discrepancy was 3.3. Oscillating neutrinos, even if confirmed, seemed not likely to be the whole answer.

While all the flap over the Reines experiment was going on, plans were proceeding for installation of the first stages of an entirely new kind of neutrino detector in the Homestake mine.

It, unlike the chlorine-37, would be able to detect the abundant but weak neutrinos produced in the sun's main proton-proton reaction. It would consist of a solid block of a metal called gallium-71. Solar neutrinos of even relatively low energy can convert gallium atoms into germanium-71, which is radioactive and detectable by techniques of radiochemistry.

Scientists from the University of Pennsylvania, the Institute for Advanced Study in Princeton, the Max Planck Institute in West Germany, and the Weizmann Institute in Israel were collaborating with Davis's group in Brookhaven on the project. The U.S. Department of Energy and the Federal Republic of Germany supplied funding for the prototype detector containing 1.5 tons of gallium. The idea is to add identical 1.5-ton blocks to it as more gallium becomes available. The scientists estimated that 50 tons of gallium-71 would be required for a full-scale experiment capable of detecting one solar neutrino per day.

Gallium is hardly an everyday substance like cleaning fluid. It is very expensive. Only in recent years has it come into widespread use, mainly for making light-emitting diodes for readout displays on calculators and in other electronics applications. Fifty tons of it would cost $25 million. But it wouldn't be harmed in any way by the experiment and could be resold later, perhaps even at a profit!

There have also been suggestions to use the element indium as a neutrino detector. When the nucleus of indium-115 captures a neutrino, one of its neutrons converts to a proton, and it becomes tin-115. The tin-115 atoms are momentarily in an excited (especially energetic) state, and when they return to normal they each emit two gamma rays, whose flash would be the sign of the neutrino's passage. The reaction also releases an electron, and its energy would depend upon the energy of the neutrino that hit. So this might make it possible not only to detect the presence of solar neutrinos but also to measure their various energies.

From the center of the sun to the shores of the Savannah River, from the chlorine in the Homestake gold mine to blocks of gallium in the next-generation neutrino observatory, the search for the final answer to the solar neutrino mystery continues. It has been a strange odyssey, to be sure, fitting perhaps for such a strange, ephemeral particle. But the expectation of an eventual certain solution to one of the truly challenging and significant scientific puzzles of recent decades makes it all worth it.

5. THE SHAKING SUN

When we watch a star, it seems to shimmer or vibrate. That is an effect of the earth's atmosphere and has nothing to do with the star itself. But in the past few years we have discovered that our own star, the sun, literally does shake or vibrate, over a dazzling range of ways, like a bell being rung.

These solar oscillations are small and difficult to detect. They have stirred all kinds of controversy. Many scientists have felt the reported observations weren't revealing any real oscillation of the sun itself. But in 1979 and 1980 a variety of new supporting evidence emerged, and the oscillations now look more certain than ever. It does appear that our sun oscillates—over periods as short as five minutes to as long as two hours forty minutes. The very material of the sun is pulsating.

These discoveries are intriguing in their own right. They add another new item to our list of previously unsuspected ways the sun is changing. Some of the oscillations are upsetting certain theories about the sun. But scientists are even more excited about their discovery for another reason: the oscillations carry information about the inside of the sun. They originate at different levels in the sun, some from just beneath the surface and some from the deep interior. Their properties vary according to the conditions they pass through inside the sun.

Just as a geophysicist studies seismic waves that travel through the earth to understand our planet's interior, solar scientists plan to study the waves traveling through the sun to understand its interior structure. A new science of "solar seismology" has just been born. It promises to be a powerful diagnostic tool allowing us to see inside the sun. One group of scientists calls solar seismology perhaps "the most exciting advance" among all the variety of new observational tests now emerging that "will revolutionize our ability to understand the sun."

Why should the sun shake? It's not altogether surprising. All mechanical bodies—the sun is no exception—can oscillate with a variety of frequencies. Pulsations produced can be of two kinds. Which kind depends upon the nature of the force that maintains the oscillations.

The first is identical to a sound wave. Called a pressure, or p-mode, wave or sometimes an acoustic wave, it consists of alternate compressions and rarefactions propagating with the velocity of sound. When you strike the end of a steel bar, you produce a pressure wave.

The second kind is called a gravity, or g-mode, wave. Gravity waves are created when a parcel of fluid (the solar gas is a fluid) is displaced from its normal position, and gravity provides the restoring force. They require a variation in the material's density with depth. Ocean waves are the most common example of gravity oscillations, but such oscillations can be seen in many circumstances.

In a sense the sun is like a complex musical instrument, only one vibrating in an incredible number of different ways as well as at a variety of frequencies. The sun is spherical, and vibrations can propagate through the body of the sun or around its circumference at some depth. From its surface to its center, the sun's temperature increases by about 2000 times and its density increases by about 1 billion times. So a single type of oscillation can manifest itself in many complex ways. There can be several nodes between the center and the surface and many nodes around the circumference of the sun.

What is it that oscillates? The physical gaseous material of the sun itself is displaced slightly and then restored in location at regular intervals, back and forth like the water in a tide pool. Temperatures, pressures, and therefore densities of the solar material fluctuate. The oscillations can be detected variously as a small rising and falling of the solar surface measured at the center of the sun's disk or as tiny fluctuation in brightness at the edge of the sun caused when the oscillation energy is dumped there.

The simplest of these apparent oscillations, occurring about once an hour, Robin Stebbins of the Sacramento Peak Observatory in New Mexico calls the "breathing mode," an in-and-out oscillation of the entire sun. The sun can also expand in one or more regions while simultaneously contracting in others, like a football being squeezed first from its ends and then from its seams. The sun can be vibrating in many such modes simultaneously. Stebbins estimates that there may be 10 million such modes in the sun. "There is no reason to believe any are excluded."

Oscillatory motions on the sun were first discovered in the early 1960s. There were the so-called five-minute oscillations, and they have been studied in great detail ever since. Throughout most of this time, solar scientists thought they were simply a local phenomenon of the sun's atmosphere. Only at the end of the 1970s were some of them shown to be oscillations of the entire sun. There have been other recent developments with these five-minute vibrations too. But we'll get back to them later.

Global solar oscillations first caught the attention of solar astronomers in 1975. That is when Henry A. Hill, an astrophysicist at the University of Arizona, and Stebbins, who was then at Arizona, interpreted long-period oscillations they were seeing on the sun as global.

Were they originally looking for oscillations? You might have guessed. This was another case in science of finding one thing while looking for another. They were carrying out precise measurements of the solar diameter in both the east-west and north-south directions to try to determine whether the sun is oblate, or slightly flattened at the poles. Oblateness is an important test of two rival theories of gravitation. Einstein's general theory of relativity successfully predicted a previously unexplained slight advance of the planet Mercury's orbit around the sun by just the amount observed. But a slightly different theory of gravitation proposed by the Princeton University physicist Robert Dicke and colleague Carl Brans suggested that if the sun were out of round by only about 1 part in 40 million, Mercury's orbit should advance less than Einstein's theory had predicted. Thus Einstein's general theory of relativity would be wrong.

Dicke had carried out detailed measurements of the sun in the 1960s and thought he saw the oblateness. But the observations were much disputed. He was actually measuring not oblateness but excess brightness in different parts of the sun's disk, and the findings could be explained away by the sun being slightly brighter at the equator than at the poles. Since certain bright solar features cluster at the equator instead of the poles there was reason to believe that might be the case.

In the 1970s Hill, who had built much of Dicke's original equipment in Princeton, designed a new kind of apparatus and set it up at an observing station 8500 feet up in the Santa Catalina mountains northeast of Tucson. The facility was called SCLERA, for Santa Catalina Laboratories for Experimental Relativity by Astrometry. The equipment measures the apparent solar diameter by detecting the edge of the sun simultaneously at opposite sides of the sun.

The oblateness sought by Dicke was not there. But Hill and Stebbins saw something else in their data. It wasn't obvious at first. They gave a paper in July 1974 in Tel Aviv reporting the results of their work on Dicke's oblateness theory. The data in that paper also contained some perplexing oscillations in their measurements of the sun's apparent diameter. Over the next few weeks Hill spent considerable time puzzling over that problem. It looked at first like merely what scientists called a "noisy signal"—random fluctuations in the measurements probably having nothing to do with the sun itself.

"I had been wrestling with my data," Hill recalled recently. "I woke up one morning about four o'clock thinking about this." That in itself wasn't unusual. Hill frequently awakes very early and does his most creative thinking in the two or three hours before getting up in the morning. "I said to myself that if we had global oscillations of the sun, that would explain the noise." Hill was so excited by the realization that he wanted to call his colleague Stebbins right then. But he managed to contain his exuberance for the next hour and a half, while sitting impatiently at his desk, until 5:30 A.M., when he knew Stebbins would be beginning observations of the newly risen sun at the solar telescope.

What the data were telling them, Hill had concluded, was that the sun itself was oscillating. There were fluctuations in the sun's apparent diameter (later interpreted as fluctuations in the brightness at the edge of the sun). On close examination the fluctuations were shown to be at regular intervals. They could be explained as the result of solar oscillations, waves coursing through the sun as though it had been rung like a bell. Checking all their data carefully, they saw oscillations at several frequencies simultaneously, most of them repeating at intervals ranging from ten minutes to about sixty minutes.

At a Texas Symposium on Relativistic Astrophysics in December of 1974, Hill and Stebbins announced their conclusion that they had observed global oscillations of the sun.

That view was quickly disputed. Scientists using other observational techniques looked for these long-period oscillations and couldn't find them. They searched for slight variations in velocity toward and from the earth of material at the center of the solar disk (velocities of several meters per second might be expected), or slight changes in temperature the oscillations should bring about, without success. Hill and his colleagues responded by pointing out that just how these matters should be affected by oscillations is difficult to predict.

Others contended that the reported oscillations were not statistically significant. By this they meant that the ups and downs recorded may not be due to any series of regular oscillations but are only what scientists call "random noise." Later analysis has shown that to be unlikely.

Others felt that the oscillations may be an effect of the earth's atmosphere and have nothing whatsoever to do with the sun. Hill and colleague Thomas P. Caudell examined this question carefully. They reported in 1979 that such differential refractive effects in the earth's atmosphere can be ruled out as an explanation of the observed oscillations.

Concluded Hill and Caudell: "Global oscillations of the sun are being observed." Acoustic and gravity oscillations in the sun itself "are the only viable mechanisms capable of producing these phenomena."

That didn't settle the matter. At least two questions remained. Were the observations repeatable, and do the oscillating waves remain in phase over many days? If the answer to either were negative the case for reality of global solar oscillations would be weakened considerably. Caudell, Hill and coworkers Jamie Knapp and Jerry D. Logan took a new set of solar oscillation measurements in 1978 and analyzed them carefully with these questions in mind. During a 23-day period that spring they were able to make observations on 18 days, digitally recording a new diameter measurement every 8 seconds.

They compared the pattern of frequencies of the oscillations obtained in these new measurements with the pattern from a similar series made in Arizona with their equipment in 1975 by Timothy M. Brown of the High Altitude Observatory and University of Colorado in Boulder (Stebbins and Hill were coauthors of that work). At a meeting on stellar pulsations held in Tucson in 1979, Caudell, Hill, and colleagues reported the results of their analysis. Twenty of the twenty-nine peaks in the 1975 records showed up in the 1978 measurements.

Was this degree of correlation significant? They found it was. "It is statistically a good indication of the repeatability of the phenomenon in nature." Given all the ways signs of oscillations can become fuzzy in the observational records and the fact that some of the characteristics in the sun responsible for particular oscillations are likely to change over three years, the correlation looked even better. The agreement between the 1978 and 1975 frequency patterns, they declared, "must be considered good." Repeatability had been demonstrated.

An even more crucial test was whether the oscillations remained in phase over a period of time. If they quickly went out of phase, it would be an indication that what was being seen was only random fluctuations, not global oscillations of the sun. Coherency of phase had been found by Caudell and Hill in one previous study. They now carried out a similar phase analysis with the new observations. As they reported at the Tucson meeting: "A considerably more convincing result is found." They acknowledged that it can be very difficult to estimate probabilities concerning the statistical significance of natural phenomena. The probabilities calculated depend on assumptions made. They got around this problem with a more direct approach of numerical simulation using large amounts of time on the computer.

This analysis showed that of twelve different oscillations observed, *each* was indeed staying in phase over the twenty-three-day span the measurements were taken. The chance that *any one* of the patterns seen could have been produced randomly was less than one in a thousand. "This phase data constitutes one of the strongest pieces of evidence for the global nature of the solar oscillations to date," they reported to their fellow astronomers and physicists at the meeting. "In summary, the new set of solar diameter measurements made at SCLERA confirms the existence and repeatability of the solar oscillations and lends strong evidence to their global nature."

Did others agree? During the 1979 Tucson meeting, Hill and his colleagues gave all the phase data they had analyzed to Douglas

Gough, a theoretical physicist involved in solar oscillation studies at Cambridge University in England. Gough chose a sample of the data at random to analyze himself.

He arranged the information on a chart in a certain way. If straight diagonal lines could be drawn through the points representing the positions of the wave fronts day by day, that would be a strong case that the oscillations were remaining in phase throughout the more-than-three-week period. If diagonal lines had to be crooked to connect the points, phase coherency was not being maintained and the reality of the oscillations would come into question.

It was a connect-the-dots game of childhood, but with a serious purpose: to reveal an underlying pattern possibly occurring in nature. "In other words," explained Gough, "if the points on the diagram represent the positions of trees, does the diagram resemble more closely a map of an orange plantation or a map of a natural forest?" He carried out his exercises. He found order, not randomness. The points were arranged in diagonal rows. "Thus one is forced to conclude that the figure resembles an orange plantation, albeit poorly laid out, rather than a forest, and that the diameter measurements do, therefore, maintain phase."

These studies had turned the tide.

"I think the scientific community now feels compelled to believe that the sun has global oscillations," Hill told me in the fall of 1980. "I think that the majority of workers realize now that they must be real."

While the scientific community had been debating the reality of Hill's long-period oscillations, scientists from the Crimean Astrophysical Observatory in the Soviet Union found signs of even longer pulsations. In January of 1976, Valerie A. Kotov, A. B. Severny, and T. T. Tsap reported in the English journal *Nature* that they had observed evidence that the sun's surface was rising and falling by a small amount in a cycle lasting two hours forty minutes.* No pulsations of this length had ever been seen previously. But the possibility was exciting because the slower the oscillation the deeper its origin

---

In the same issue of Nature, *a group from the University of Birmingham, England, also reported detecting the oscillations. Such simultaneous publication usually means that the two groups share the honor of discovery. However, the same-issue journal publication was made possible in this case only because the Soviet paper was, for a variety of reasons including translation difficulties, delayed while the Birmingham paper was published especially quickly. In fact, the Soviet scientists' paper had been submitted nearly five months before the English group's.*

within the sun, and the more it should eventually be able to reveal about the sun's deep interior.

Once again questions arose as to whether the perceived oscillation was something real happening on the sun and whether it could be observed by other investigators elsewhere. The Soviet group reported that the rate of movement was quite small, only about one meter per second. That strained the observing technology to the limit.

At Stanford University, solar physicists Philip H. Scherrer and John M. Wilcox quickly began a program to look for the 160-minute oscillations. Similar methods were used at both the Crimean observatory and the Stanford Solar Observatory. The scientists measured the difference in line-of-sight velocity between a central area of the sun's disk and a surrounding ring-shaped area. These motions toward or from the earth cause slight shifts in the wavelength of the light, the so-called Doppler effect, and these shifts are what is detected. The same effect is responsible for the varying pitch of a train or auto horn as it first approaches and then moves away from you.

Scherrer and Wilcox joined into collaboration with the Soviet observers. The patient efforts of Kotov, who spent four weeks at Stanford, helped iron out some early difficulties in the data analysis. By 1979, after three years of work, the scientists were able to report that the observations at Stanford appeared to support the results first reported in the Crimea. The Stanford scientists had observed the sun with their equipment for several hundred hours during each of the previous three summers, and they not only saw the oscillations but found that they were in phase with the observations in the Crimea. That was strong support for their reality. If both groups were observing a solar phenomenon they would expect to see them in phase. In other words, when the start of a new oscillation was seen at Stanford it was also seen at the Soviet observatory. If it were only some terrestrial phenomena that was being seen, or the effects were due to "noise," no such phase agreement should be seen. Said Scherrer at the spring meeting of the American Geophysical Union in Washington in 1979: "We do tend at this time to support the reality of these observations."

The observations were continued in 1979. It was slow going at first because of bad weather (too many cloudy days) in northern California. But by the end of the year the U.S.-Soviet collaborators had compiled 227 more hours of sun-watching at Stanford and 246 hours in the Crimea.

These new observations helped eliminate one of the most

troublesome concerns about the data. The oscillations were about 160 minutes in length, and 160 minutes is exactly one ninth of a day. Now that is disturbing. Any fluctuation that is an exact fraction or multiple of the length of the day is suspect. It opens the possibility that the observations are telling us something about the earth or its atmosphere rather than the sun.

"If the oscillations were exactly one ninth of a day I don't think we'd ever believe it was real," Scherrer said. "It makes you say this is probably not from the sun, this is probably from the earth's atmosphere. The sun doesn't know how long a day is, and if your data say something about the length of a day then you are probably not measuring the sun."

But now there were enough observations to say with certainty that the oscillations *weren't* exactly 160 minutes in duration. They were slightly longer. They were, to be precise, 160.01 minutes long. The difference is admittedly small, less than a second. But to the observers at least, it was enough to rid them of the one-ninth-day criticism. The scientists also now had four years of data at two widely separated locations with all the wave patterns still in phase. "It is very difficult," they reported in May 1980, "to conceive of any observational or statistical artifact that could produce such close agreement of phase with a period that is not exactly one-ninth of a day.... The continued agreement in phase (and amplitude) between the two observatories, as well as the fact that the period differs from exactly one-ninth of a day, supports the interpretation that we are indeed observing solar oscillations."

But there was one more troubling point. A group of French scientists, concerned that 160-minute oscillations aren't expected from our standard models of the sun, looked for them with their own equipment at the University of Nice. They didn't see them. That threw everything into confusion.

The issue was so important that the French, Soviet, and American investigators decided to collaborate. One of the big problems in studying these slow solar oscillations is the difficulty of getting long sequences of observing time under clear and consistent sky conditions. One place where that problem can be averted is at the South Pole. There in the Antarctic summer, the sun remains in the sky twenty-four hours a day and it remains at essentially the same elevation about the horizon, making a complete circle of the sky. It is quite eerie, in fact. I have been there, and it is astonishing to go outside at two in the morning into broad daylight, the sun shining brightly.

These advantages and the fact that the summer Antarctic weather is usually good for several days at a time led the French scientists, Eric Fossat and Gérard Grec, to Antarctica. There they worked with an American scientist, Martin Pomerantz, of the Bartol Research Foundation of Philadelphia's Franklin Institute, which has two observatories in the Antarctic. For five days they made continuous observations with a telescope five miles from the United States' domed research station at the South Pole. These five days at the Pole yielded almost as much data as Stanford and the Soviets had been able to collect in an entire year.

The French scientists then took their data to the Crimea for analysis. The results were exciting. "A couple of weeks ago we got a telegram from them," Wilcox announced from Stanford on July 31, 1980. "It said: '160 minute oscillation is present in South Pole data. The amplitude is 33 centimeters per second and phase is in perfect agreement with yours.'"

The oscillations were undoubtedly real.

Why hadn't the French scientists been able to see them in the observations at Nice? Wilcox and Scherrer speculate that it was probably because of atmospheric turbulence over the Riviera. Even on clear days there the air is turbulent enough to block the small Doppler shifts in wavelength of the light being measured. The actual motions of the sun, as measured at the South Pole, are only a little over a foot per second, so it is hardly surprising that they couldn't be seen with the less than excellent conditions in France.

The South Pole observations convinced all the scientists involved in the collaboration that the 160-minute solar pulsations were valid. Said Wilcox: "The theorists in the scientific community, including ourselves for that matter, will really have to take this seriously."

Throughout the late 1970s nearly all the attention concerning solar oscillations was being directed to the long-period observations of Hill and of the Stanford and Crimean groups. The five-minute oscillations known since the early 1960s had seemingly been almost forgotten. They had been studied in some detail, but they were thought to be fluctuations in local areas of the sun They seemed to have little in common with the longer-period oscillations, which course throughout the entire sun. The five-minute vibrations were generally considered to be due to movements of randomly distributed, independent cells of solar materials perhaps 5,000 to 10,000 kilometers across and lasting only five to ten fluctuations, about half an hour to an hour.

New observations made at the Sacramento Peak Observatory in New Mexico from 1975 to 1978 changed all that. The observatory is 9,200 feet up in the conifer-clad Sacramento Mountains of south central New Mexico, and it is one of the nation's premier solar observing facilities. (Other excellent ones are on Kitt Peak in Arizona and Mount Wilson in California.) Its Vacuum Tower solar telescope makes use of a huge vertical shaft 315 feet long (more than half of it beneath the ground) from which all air has been evacuated. This vacuum allows the sunlight to reach the recording instruments without being disturbed by air motions inside the telescope. The result is observations of the sun of a resolution previously possible only from a space satellite or a high-altitude balloon.

Franz Deubner, a Sacramento Peak visiting scientist from West Germany, Roger Ulrich of UCLA, and E. J. Rhodes, Jr., of Caltech used an array of photoelectric recording instruments attached to the telescope to monitor the five-minute oscillations. They registered the

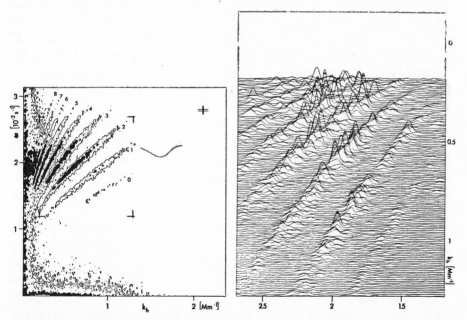

Left: Plot of eight different solar "p-mode" oscillations. Right: Solar p-mode oscillation data plotted as "eastbound" wavetrains on sun. (Deubner, Ulrich, and Rhodes)

Doppler fluctuations due to line-of-sight velocities on the sun over a large area of the center of the sun's disk. To resolve the oscillations in the unprecedented detail they did placed large demands on the telescope's pointing accuracy and on the computer's data-handling capacity. Thirteen million measurements of line-of-sight velocity were made in seven hours during one such observing period in 1978. The results were well worth all the trouble.

The observations showed that the five-minute oscillations are not local events but a global phenomenon of the entire sun. These short-term vibrations pervade the sun's convection zone and remain very much in phase in space and time. A typical horizontal wave travels around the sun at about fifty miles per second, completing a circumnavigation of the sun in a little under fifteen hours.

The work meant that the five-minute oscillations, like the longer-period ones, carry an enormous amount of information about the solar interior. "This discovery," said Sacramento Peak physicist Lawrence Cram, "is one of the most important advances in solar physics." Deubner and colleagues also noted the potential. "Identification of the five-minute oscillations as a global phenomenon . . . has given us a tool to observe directly systematic large-scale motions in the solar interior," they wrote in the 1979 paper describing their results. An extension of the work to the whole disk of the sun "will eventually provide a quantitative three-dimensional picture of the large-scale circulation pattern of the solar interior."

Physically, the five-minute oscillations are a standing pattern of sound waves trapped within the sun's convection zone. This outer thickness of the sun is where all the sun's magnetic fields are generated and thus where the solar cycle is regulated. So understanding it is crucially important to better understanding solar activity.

Analysis of the five-minute oscillations has already helped place certain limits on the depth of the zone. The convection zone seems to extend down at least a fourth and perhaps as much as a third of the way to the center of the sun (although considerable controversy surrounds this conclusion). This is about 50 percent deeper than previous estimates. The work by Deubner also shows that the pattern of oscillations on the solar surface can be used as "tracers" that will help solve another important question about the zone: the variation of solar rotation with depth. Previous studies of sunspots indicate that the level where they and other magnetic features are "anchored" or generated rotates faster than the sun's surface. The new studies using the surface pattern of five-minute oscillations as tracers show the same thing. Even relatively shallow layers of the sun, six to ten thousand miles down, seem to rotate faster than the surface. Such work, the scientists think, should eventually provide a detailed profile of the sun's all-important convection zone: its depth, its structure, horizontal currents within it, and its varying rotation speed with depth. And all that should reveal much about the workings and variations of the solar dynamo, which is responsible for all solar activity and its changes.

The oscillations with periods longer than five minutes allow us to perceive even deeper into the sun. While most of the five-minute vibrations are pressure waves traveling through the outer, convection zone, the low-frequency waves move through deeper layers of the sun. They are probes of the deep interior, messengers from the far inside of the sun.

They can tell us about conditions there. One oscillation, for instance, occurs at a period of about forty-eight minutes. Calculations show that the frequency of this oscillation would be different by about 2 percent if the sun's interior is mixed instead of neatly stratified, as most theories assume. Precise measurements of this oscillation over a long enough time can distinguish between these two possibilities.

Robin Stebbins took a newly designed telescope to the South Pole during the 1980–81 Antarctic summer to begin the kinds of continuous observation scientists need to sort out all these oscillations periods of the sun. From the South Pole, from widely spaced sites around the earth, and eventually from space, solar scientists hope eventually to distinguish all the different kinds of solar vibrations. This fine-scale resolution will allow the vibration modes of the sun to be classified in much the same way that excited states of the atom and of the nuclei of atoms are now classified. All these distinctions should give us unprecedented information about the inside of the sun.

So at last solar scientists have almost in their grasp a tool equivalent to the medical diagnostician's X rays of the human body or the seismologist's seismic profile of the earth's insides. The subject is in its infancy. Yet the potential is clear. The sun's own vibrations will reveal its inner secrets, if only the human observers 150 million kilometers away can be clever enough to decipher the codes and read and interpret the information they contain. The shaking sun is a revelatory sun.

6. THE SHRINKING SUN

Scientists contemplating the sun in the mid-nineteenth century were in a quandary. Geologists had shown that the age of the earth had to be measured in millions of years, not thousands. Yet it could be easily calculated that a sun powered by chemical burning could never last that long. For instance, Sir William Thomson (who later became Lord Kelvin) had shown that if the sun were composed of solid coal and produced its heat by combustion, it would burn out in less than five thousand years.

What then could account for the sun's energy? Nuclear energy would not be known until well into the twentieth century. Well, two sources seemed possible. Both concerned gravitation.

When an object falls to the ground, its energy of motion is not lost but transferred—much of it into heat. Now if meteors flashing through space were drawn into the sun by its gravity in great enough numbers, the energy they transferred to it could be enough to provide the sun's heat. One calculation showed that a quantity of matter equal to only about one one hundredth the mass of the earth falling annually into the sun would be enough to maintain its radiation indefinitely.

There were, naturally, problems with that idea. If meteoric matter were so abundant, earth ought to encounter far more of it than it does. And such an infall into the sun implied even vaster quantities of matter in the vicinity of the sun. The presence of such masses should produce detectable effects on the motions of the innermost planet, Mercury—effects that were not seen.

But there was another source of matter—the sun's own mass. Suppose that due to the force of its own gravity, the sun were gradually shrinking. Matter in its outer layers would be falling in onto matter slightly closer to the center. The collisions would produce heat, in abundant quantities. This idea was worked out by Hermann Ludwig Ferdinand von Helmholtz, a brilliant German physician, physiologist, and physicist (those were the days before the tyranny of narrow specialization).

He showed that the process of gravitational contraction was sufficient to heat the sun to a temperature of 28 million degrees F. (Amazingly, this is almost exactly the temperature we now believe the interior of the sun is at.) The sun would acquire "a store of heat not only of considerable, but even of colossal, magnitude."

Previous condensation would have been sufficient to have powered the sun "for not less than 22 million years of the past," Helmholtz figured. And the process could easily continue long into the future. "This would develop fresh quantities of heat, which would be sufficient to maintain for an additional 17 million of years the same intensity of sunshine as that which is now the source of all terrestrial life."

Helmholtz first proposed this concept of a sun powered by its own slow shrinkage in a popular scientific lecture in 1854. It was an eminently reasonable idea, even though it didn't give geologists all the time they thought necessary to explain the past evolution of the earth. It soon gained wide scientific acceptance, and it dominated scientific thought on the subject until the twentieth century.

But wouldn't the shrinkage be immediately obvious? Not at all. The necessary contraction in the sun's diameter would be imperceptible. Take, for example, this discussion of the matter I found in an 1895 textbook on the sun by astronomer C. A. Young of Princeton University:

"...A contraction in the sun's diameter of about three hundred feet a year—a mile in a little more than seventeen years—would account for its whole annual heat-emission. This contraction is so slow that it would be quite imperceptible to observation. It would require seven thousand years to reduce the diameter a single second of arc, and nothing less would be certainly detectable."

So the sun was assumed to be slowly shrinking. That view held sway until physicists developed the theory of nuclear generation in the sun in the 1920s. Gravitational contraction was no longer necessary.

The idea of a shrinking sun was forgotten—until 1979. That's when John A. Eddy, the solar astronomer we met in chapter 3, resurrected the idea. It was reborn not out of any great theoretical need but because he found that past observational records of the sun did seem to show a shrinkage.

The work, Eddy says, was an outgrowth of his inquiries into the Maunder sunspot minimum. In the course of those investigations, he had learned of a remarkable series of observations of the sun taken almost daily at the Greenwich Observatory in England for two hundred years.

The observations are made by what astronomers call a meridian transit telescope. The Greenwich telescope's axis, by international agreement, defines the line on earth of 0° longitude, the prime meridian. The telescope was used to observe and time the exact instant the sun crosses the prime meridian at noon. This is a measurement used for determining the sun's exact location against the background of the sky, which is essential for precise determinations of earth's orbit.

The telescope can swing up and down vertically but not horizontally. It views the sky from a slit in the south side of the building. To find when the center of the sun crosses the meridian, the observer records first the moment of the passage of the right, or leading, edge across that line and then the left, or trailing, edge. The center's passage is the midpoint. The first such telescope was installed at Greenwich in 1750, and the measurements carried out daily (except for cloudy days) ever since.

The early observers had to use an eye-and-ear method of observation. Lying on a couch beneath the telescope's eyepiece, they would look at the pendulum clock as the sun's edge approached the telescope's crosshairs and then listen for the number of tick-tocks until the moment it crossed them. Needless to say, accuracy to a tenth of a second was difficult. But in 1850 a better telescope and in 1854 a chronograph for better timing were installed. From that point on until 1953, when the observatory was moved to a location farther south, the records were quite good, in Eddy's view. They provided one hundred years of data from the same instrument.

Now, the difference in time of passage of the right and left edges of the sun is obviously a measure of the sun's horizontal diameter. What Eddy and mathematician colleague Aran A. Boornazian saw when they examined the Greenwich records surprised them.

There are annual variations in apparent diameter due to the fact that the earth, with its elliptical orbit, is slightly closer to the sun in the winter and slightly farther away in the summer. These are huge fluctuations and easy to subtract out of the records. When those changes and other known orbital effects were eliminated, a declining trend in the diameter was clear. The sun seemed to be shrinking.

"This is what we got excited about," Eddy recalls. "We saw a gradual downward trend. Our first impression was that it wasn't real." But then he discovered that the U.S. Naval Observatory had been doing the same kind of observations, and their records also showed the decline. "When Greenwich and the Naval Observatory saw a shrinking trend that was pretty much the same, that seemed to reduce the arguments against it."

The amount of apparent shrinkage was extraordinary. It amounted to an average contraction of 2 seconds of arc per century.

At the distance of the sun, an angular width of 2 seconds of arc is equal to 1550 kilometers, or 960 miles. So, incredibly, the sun seemed to be shrinking nearly 10 miles a year, or nearly 6 feet per hour.

Now, the average angular width of the entire sun as seen from earth is a little over half a degree. To be exact, it's 31'59". That is 1919 seconds of arc. If the sun were to continue shrinking at 2 seconds of arc per century, it would totally disappear in 960 centuries, or only another 96,000 years.

No one, least of all Eddy and Boornazian, seriously believed that was going to happen.

What seemed more likely was that, if the effect is real, the sun goes through cycles in which its diameter fluctuates. In that event, these measurements happened to have been made during the shrinking phase of that cycle. The shrinkage was assumed to be taking place only in the sun's outer layers, not the entire solar mass. Were it otherwise, the shrinkage would have produced more than two hundred times the observed luminosity of the sun!

There was some other evidence interpreted as supportive of a shrinking sun. In 1567 the path of a solar eclipse passed over Rome. Calculations show that the moon was then at a point near enough the earth that all of the sun's light should have been blacked out. It wasn't. The astronomer Christopher Clavius, a contemporary and teacher of Galileo, wrote that the moon "did not obscure the whole Sun . . . but a certain narrow circle was left on the Sun, surrounding the whole of the Moon on all sides." Such an annular ring could be explained if the sun were larger four centuries ago than it is now.

There was one troubling aspect to the observations. The Greenwich observers had also measured each day the solar diameter in the vertical direction. These records showed a shrinkage too, but only by half the amount measured horizontally. If the sun were really shrinking, why would it do so more in one direction than another? Eddy felt the discrepancy had an observational explanation. The vertical measurement was performed by swinging the telescope vertically to the sun's upper and lower limbs as its disk crossed the telescope. This was done during the two minutes the observer had between right- and left-limb passage.

He assumed the horizontal measurement should be the more accurate because it involved no motion of the telescope and thus no potentially deflecting wear and tear on the adjusting machine screws. He says it wasn't until later, when he visited the Royal Observatory and actually performed the observation himself, that he found otherwise. The vertical measurement could well be the more accurate, he discovered, because the observer has more time to decide on the exact position of the sun's upper and lower edge in the telescope's crosshairs. In contrast, the horizontal-diameter measurement requires making a very quick judgment about exactly when the edge of the moving sun crosses the telescope's center. So the more reliable data showed only half the maximum amount of shrinkage.

By this time, however, Eddy had already reported his original conclusions to the American Astronomical Society, in a meeting at Wellesley, Massachusetts, in June 1979. The sun's diameter, he concluded, was apparently variable. Still another changing aspect of the sun had been discovered.

It was an intriguing possibility. Here was another totally

surprising discovery about the sun. Beyond that, it might explain the great missing neutrino mystery (chapter 4). If the sun was contracting, some of its heat would be generated by that means, a revival of Helmholtz's idea a century and a quarter earlier. That would mean that a lower rate of reactions in the sun's thermonuclear core—and fewer neutrinos—could still account for the total energy output seen.

Said Eddy and Boornazian in their paper to the assembled astronomers: "If only the outer 20 percent of the sun's radius is involved—the convective zone—enough energy would be supplied to make up the deficit that falls when we take the presently measured neutrino flux as indicative of the real temperature of the solar core. The implication is that the sun and presumably other similar stars could now be deriving a significant part of their energy from gravitational contraction."

No sooner had they given their report than, Eddy recalls good-naturedly, "we started getting shot at." Many fellow astronomers were dubious.

Almost the same week, a group of scientists led by Sabatino Sofia and John O'Keefe from NASA's Goddard Space Flight Center in Maryland reported in the journal *Science* an amount of solar shrinkage much less than what Eddy and Boornazian had found. Sofia and his colleagues chose to analyze only the Greenwich vertical-diameter data. They considered the vertical measurement more reliable because it was done with a micrometer, not a clock. They found evidence for a slow systematic decrease of the observed radius by about 0.2 arc second between 1850 and 1937. The rate seemed to be only about one fifth to one tenth the amount Eddy had found.

That fall, at a conference on the ancient sun, held in Boulder, David W. Dunham, an astronomer with a special interest in the timing and measurement of celestial events, reported results of an intriguing new measurement of changing solar diameter. It made use of eclipse observations. Dunham and his colleagues had stationed observers at the edge of the path of totality of the February 26, 1979, total solar eclipse in the Pacific Northwest. They made careful observations of the exact time of totality at each location. From them the solar diameter could be accurately calculated. Then Dunham compared their observations with those made by the British astronomer Edmund Halley during an eclipse in 1715. Halley had arranged for hundreds of observers to be strung out over the path of totality across England. The purpose was not to measure the solar diameter but to see how well Halley had predicted the path of totality. But from the records of these observations, Dunham was able to calculate the solar diameter for the year 1715. He found it to be about 0.34 arc second larger than in 1979. This amounted to about 0.13 arc second per century shrinkage.

Again, the amount of change was far less than Eddy had found. But some shrinkage was indicated. Eddy was pleased. "It suggests that the solar diameter *is* changing," he commented at the end of Dunham's presentation. "I think that's exciting, although it doesn't really confirm our two-seconds-of-arc-per-century figure." By then he had begun to realize that he and Boornazian had indeed placed too much confidence in the accuracy of the Greenwich horizontal-diameter measurements.

Other types of measurements soon came forth. At the Mount Wilson Observatory in California, astronomer Robert F. Howard reported that his analysis of the solar diameter using magnetograms of the sun made daily with a photoelectric scanning device since 1968 revealed noticeable shrinkage. The contraction during the following ten-year period amounted to a rate of 2 arc seconds per century. An examination of photos of the sun from 1930 to 1980 revealed a smaller rate of 0.5 arc second per century.

So the sun definitely seemed to be shrinking, but the exact amount was in great dispute. Then came Irwin I. Shapiro with his data on transits of Mercury across the sun's disk.

About thirteen times a century, Mercury, the innermost planet of the solar system, passes in front of the sun, as seen from earth. These transits are measured very carefully because knowledge of the exact orbit of Mercury is essential to verification of modern theories of gravitation. Shapiro, a scientist at the Massachusetts Institute of Technology, has been involved in this work, and so he had all the data handy.

Shapiro realized that any shrinkage of the sun of the amount Eddy had reported would have enormous importance to our understanding of the sun. So he decided to analyze his computerized data of past transits of Mercury to see if they revealed any change in solar diameter. Any shrinkage should show up as a slight reduction in the time Mercury takes to cross the sun's disk (about five hours). He examined the records of twenty-three such transits of Mercury between 1736 and 1973.

In April 1980 Shapiro reported his results in *Science*: "No statistically significant change in the diameter of the sun." Actually, the Mercury observations weren't precise enough to rule out any change in the solar diameter. The observations had large uncertainties. But they did show that any shrinkage had to be less than 0.3 arc second per century. English scientists Leslie Morrison and John Parkinson chimed in with a similar conclusion based on their own analysis of both Mercury transit data and eclipse observations. Later in the year, however, several new studies were reported supporting some amount of shrinkage.

So what is going on? Is the sun shrinking or not? And if so, by how much? We first should remember that science does not always progress along a crisp, clearly marked path. In fact, it rarely does. Every step forward is marked by confusion and seeming contradiction. Different methods of measurement give slightly different answers. Instruments vary. So do the judgments and interpretations of human observers. This all takes a while to sort out. Scientific advance never happens with the surety that the textbooks portray. And that helps keep it interesting for everyone.

Eddy tried to put the matter into perspective during a discussion I had with him in Boulder in July 1980. "I feel now that we were probably wrong," he said of the inference by him and Boornazian that the sun was temporarily shrinking at the rate of 2 arc seconds per century. They had done many checks on the reliability of the data. For instance, the Greenwich records should reveal a slight out-of-phase seasonal modulation in the *apparent* solar diameter due to the fact that the times of earth's closest and farthest approaches to the sun fail to coincide with the solstices (the sun's crossing of the equator on June 21 and December 21) by about two and a half weeks. That effect was seen. "The data looked like you could trust it," Eddy recalls.

What then accounts for the problems? "Atmospheric influences were bigger than we thought," says Eddy. Earth's atmosphere is the big bugaboo to astronomers. It causes all sources of light to shimmer and dance. And that introduces uncertainties into all ground-based astronomical measurements. There are other pesky optical-perceptual effects as well. Everyone has seen striking demonstrations of optical illusions. A light-colored object on a dark background, for instance, looks bigger than a same-sized dark object on a light-colored background.* In the same way, the sun appears bigger to the eye against a dark sky than it does against a light-colored sky. A clear, pollution-free sky is a deep, dark blue. But as the atmosphere becomes more polluted, as it has over the past century or so, the daytime sky takes on a lighter, more whitish appearance. Perhaps, then, the gradually diminishing sun that shows up in the Greenwich and Naval Observatory (Washington, D.C.) records is only due to this perceptual effect. The sky has become lighter and so the sun appears smaller.

How can you test this idea? Well, the sky varies in brightness from horizon to overhead. Large dust particles are abundant in the lowermost part of the atmosphere. Such large particles tend to scatter all wavelengths of light equally, resulting in a white sky near the horizon when you are looking through a considerable thickness of dust-laden air. High overhead the sky is darker. There the particles that scatter sunlight are small gas molecules, and small particles preferentially scatter light of short wavelength (blue and violet). That is why the sky is blue.

In the wintertime the sun's arc across the sky is much closer to the horizon than in summer, when the sun is quite high in the sky. So if this perceptual effect I've been describing is an important influence on solar-diameter measurements, the records should also show slightly smaller diameters in the winter, when the midday sun is against the low, white sky, than in summer, when it's against the high, darker sky.

Eddy looked at the data for this effect and found it. Disregarding all orbital effects, there is a noticeable cyclical rise in the measured diameter in the summer and dropoff in the winter. So it seems likely that some of the shrinkage that appears in the records over the past hundred years is indeed only a perceptual effect due to the sky becoming more polluted and lighter-colored.

"The lightening of the sky accounts for *some* of the shrinkage we saw," says Eddy, "but not all of it." He agrees that he and Boornazian overestimated the amount of any shrinkage revealed in the solar records. "I suspect that we were wrong in our initial

---

The fact that the moon and sun appear larger than usual when rising or setting is another, but completely separate, perceptual effect due to their proximity to the horizon. But that horizon effect would not come into play in these kinds of measurements. They are made only at midday, and only the edges of the sun (and not the horizon) are observed.

assessment." Yet he points out that all the subsequent studies by other scientists have yielded results consistent with some shrinkage. He recalls the numbers that each of those studies give for a possible change in the sun's diameter. Notes Eddy, "Every one of them is negative," indicating a diminishing diameter. "Not a one is on the other side of zero."

What does he conclude from it all? "I think, yes, that the sun's diameter is changing, but how much we don't know. I think the diameter probably goes up and down slightly. And I think the convective zone itself [about the outer one fourth of the sun] is all that's changing."

It's important to find out just how much shrinkage may take place. For as first Helmholtz and later C. A. Young pointed out so long ago, even a very small amount of gravitational contraction can produce huge amounts of energy. The amount of shrinkage Eddy and Boornazian first thought they had discovered is enough to produce, by gravitational attraction, energy equivalent to 218 times the sun's present illumination. Even the amount of shrinkage not ruled out by Irwin Shapiro's analysis is enough to produce 20 times the sun's present illumination, says Eddy. So the question is obviously important to the study of possible changes in solar luminosity, the subject of our next chapter. One of the big questions is why, if the sun is shrinking, the effect hasn't yet shown up as a change in the sun's luminosity.

The final answers about the shrinking sun may be found in the 1980s. Three new programs specifically designed to make accurate, reliable measurements of the solar diameter are being started. They wil be conducted by Timothy Brown on the High Altitude Observatory's Table Mesa site overlooking Boulder, by Henry A. Hill at his observing site in the Santa Catalina Mountains northeast of Tucson, and by astronomers at Kitt Peak National Observatory southwest of Tucson.

It may be ten years before a long-enough record is compiled to see a trend in any of these precise measurements. Then we may know whether the studies of the past two years showing a small decline in solar diameter have been a fascinating but futile exercise of attempting to tease too much information out of too-imprecise sets of measurements, or whether the sun really is temporarily shrinking. Either way, the quest should be interesting and the answer intriguing.

7. IS THE SUN A VARIABLE STAR?

On August 13, 1596, David Fabricius, a Dutch pastor and the father of the man who fifteen years later would become one of the four codiscoverers of sunspots, was observing stars in the constellation Cetus, the Whale. He routinely noted a star of the third magnitude, quite easily visible. There was nothing out of the ordinary about it. Yet when he looked again a few weeks later, it had disappeared. Unfortunately, he seems not to have followed up on this oddity.

Over succeeding years it was seen again from time to time. In 1603 it was entered on a new star chart as a star of the fourth magnitude, considerably dimmer. It wasn't until another Dutch observer, Phocylides Holwarda, began a series of regular observations in 1638 that the star was shown to appear and disappear regularly, its brightness varying so that at long intervals it seemed to blink out entirely. Holwarda had made the first discovery of a genuine variable star.

This star, called Mira, is now only one of many known stars having this same odd property. Its output of light slowly rises and falls over a period of approximately 332 days. The period is not regular; maximum brightness can come as much as a month earlier or later than average. The brightness is unpredictable also. Astronomers measure the brightness of stars in units of magnitude. A difference of five magnitudes is taken as exactly a difference of 100 times in brightness. So a one-magnitude difference represents a brightness difference of 2.5. At typical maximum, Mira has a magnitude of 3.4, in the midrange of all the stars you can see in the night sky. But during some periods the red-orange Mira attains the brightness of Polaris, the Pole Star, magnitude 2. (The lower the magnitude the brighter the star.) At minimum, Mira fades from view. In fact, it is visible to the unaided eye only about 18 weeks out of its 47-week period. Magnitude 6 is about the dimmest star you can see with your eyes alone, and Mira can become as dim as magnitude 10. So its brightness can vary over a range of eight magnitudes, meaning that it can give out 1,500 times as much light at one time as at another.

It's sobering to think what life would be like on a planet around a sun as irregular as Mira. The thousands of Mira-type long-period variable stars are about as unlikely candidates for the development of life as we can imagine. Not only are they highly variable in light output, but they have enormous distended atmospheres as big as our whole inner solar system. Astronomers interested in the possibilities of extraterrestrial life don't place much hope for it around any of them.

There are other types of variable stars. Most have much shorter periods than the Mira-type variables, whose periods typically range from half a year to well over a year. There are many different kinds, but the most famous are the Cepheids. They draw their name from the first star of their type known, Delta Cephei, discovered in 1784 by the English amateur astronomer John Goodricke.

What a sight one of them would be close up! Cepheid variables swell and contract in size like a beating heart, varying in temperature and brightness as they do so. They typically go through their cycle in anywhere from two to forty days. Delta Cephei, for instance, varies in brightness by two and a half times over a period of 5.4 days. Cepheids are yellow supergiant stars, which means they are similar in color to the sun but two to five times as massive.

Cepheid variables turned out to be invaluable yardsticks for measuring the distances to other stars and galaxies. In the early part of this century, the astronomer Henrietta Leavitt, working at the Harvard College Observatory, discovered a remarkable property about them. She studied photographic plates of several small galaxies in the neighborhood of our own Milky Way galaxy, and analyzed the light curves of the Cepheid variable stars in them. She found that the brighter they appeared, the longer their periods of variation were. Now, all the Cepheids in such a star cluster should be about the same distance from us, in the same way that all the people in a town in the next state are all about the same distance from you. That means that the period of pulsation of these stars is proportional not just to their *apparent* brightness, but to their *absolute* brightness, or luminosity.

The relationship is known as the period-luminosity law, and it allowed astronomers to determine distances. If you see a light in the dark you might have no way of knowing whether it is a nearby flashlight or a far-off streetlight. But if you *do* know which it is—and therefore know its absolute brightness—then you have a good idea how far away it is. The Cepheid variables told astronomers how bright they were in absolute terms. Once the astronomers identified a Cepheid in a distant star cluster and noted its period of variation, the period-luminosity law yielded its brightness and therefore the distance of that group of stars. It actually turns out that there are two types of Cepheids—one more luminous than the other—but once the correct type is determined the astronomers have a convenient measuring rod. In the first half of this century, the application of these calculations to Cepheid variable stars told astronomers the universe was much larger than they previously thought.

A catalog compiled several years ago listed more than 25,000 variable stars, and new ones were being discovered at an accelerating rate. There is even an American Association of Variable Star Observers, whose members perform a valuable service in keeping track of them systematically.

Is the sun a variable star? Obviously not to the degree any of these 25,000 stars are. If it were, we wouldn't be here to see the show. Life probably could never have become established in such chaos and instability. Yet already in these pages we have seen a surprising degree of variation and irregularity in our star. It has magnetic-activity and sunspot cycles, and these cycles themselves are variable in intensity and somewhat irregular in occurrence, prone to sputter from time to time. The sun's neutrinos may not be as numerous as expected, or else the neutrinos it produces are unstable and change form in flight. Its diameter may go through small changes, perhaps shrinking slightly now, expanding at other times. Its period of rotation seems to have suffered some glitches in the past. The entire sun is oscillating in virtually every mode imaginable. In all these senses the sun is indeed a variable star.

When astronomers speak of the sun as a possibly variable star, however, they are referring mainly to something we haven't mentioned yet: its total output of electromagnetic radiation, its luminosity. Does *that* change? That has been a key question of solar science. It has, as we shall see, tantalized and frustrated solar astronomers. But the question of a possibly variable output of the sun has gained new urgency in the past five years with the newfound signs of all the other kinds of changes on the sun.

With so much happening on the sun to shatter our long-held view of solar regularity, isn't it natural to assume that its luminosity changes as well? It may be natural, but in science you try not to assume. You try to find out. To measure changes in solar luminosity is a dreadfully difficult task because any changes might be very small and might take place over long periods of time.

The variable stars we mentioned earlier pulsate in luminosity by anywhere from 10 percent for the most mundane ones to 500 percent for the Cepheids to 150,000 percent for an extreme variable like Mira.

We are confident now that any changes in the sun's luminosity are much smaller indeed—less than 1 percent. But the question of exactly how much it might vary is crucially important. Small changes in the amount of solar radiation reaching earth can, through feedback mechanisms on the surface and in our atmosphere, have their effects amplified. A 1 percent decrease in solar luminosity would be sufficient in itself to cause a 1°C drop in average world temperature—enough to bring a return of what's known as the Little Ice Age that afflicted Europe most strongly in the seventeenth century. Even prolonged changes as small as 0.1 percent could be enough to cause economically significant effects on climate. Solar-luminosity changes are seen as one way to explain past fluctuations in world climate and a possible means of predicting future climate change.

The total radiant energy output of the sun is usually called the solar constant. It's a misnomer because we have no real justification to assume the sun's output of light stays always the same. We are now, in fact, considering how inconstant the "constant" might be. Use of the term "solar constant" dates back to the time when solar regularity and constancy was assumed. The French scientist Claude Pouillet introduced it. In 1837 Pouillet developed the first instrument to measure the sun's total output of radiation. It was essentially a blackened box filled with water. After a time a thermometer measured the temperature of the water and the total amount of energy absorbed was calculated.

Both Pouillet and the astronomer Sir John Herschel, who developed a similar instrument, were interested primarily in trying to measure the temperature of the sun. That was beyond them because the physical laws relating a body's radiation output to its temperature hadn't been developed yet. But Pouillet's admirable efforts did yield a figure for the solar constant that was quite good for the time. It was only 9 percent below the currently accepted value.

The solar constant is now generally defined as the sun's total radiative power, summed up over all wavelengths, as measured above the earth's atmosphere, at the average sun-earth distance. The amount turns out to be about 1369 watts per square meter, with a possible error of 7 in either direction.

It's all only a technical way of referring to the amount of heat given off by the sun, on which all life on earth depends. So it is rather important to know whether it varies. First, however, it has to be measured accurately.

As you can see, all measurements of the solar constant from earth are afflicted with troublesome complications. The amount of sunlight absorbed by earth's atmosphere before it reaches the detector has to be estimated. And that is further complicated by the fact that some wavelengths are absorbed more by the atmosphere than others. Most of the sun's electromagnetic radiation is at the wavelengths our eyes see as visible light.* The atmosphere filters out some of it, but not uniformly. Like fishes in the ocean eating certain kinds of organic debris raining down through the water, ozone and oxygen molecules in the atmosphere absorb certain wavelengths of the visible sunlight. Water vapor takes out another significant bite. Some of the solar radiation is also at shorter (ultraviolet light) or at longer (infrared) wavelengths. Ozone takes a chunk out of the ultraviolet, and

---

That is not convenient coincidence. We have evolved so that our eyes see those wavelengths of light emitted most prevalently by the sun.

water vapor and carbon dioxide take serious bites out of the infrared. So all kinds of corrections have to be made for these absorptions.

Two of the real pioneers of solar constant measurements were two secretaries of the Smithsonian Institution, first Samuel Pierpont Langley and then Charles Greeley Abbot. Langley is best known in the public mind as the man who almost became the first to fly a heavier-than-air flying machine. He might have made it, too. He had worked out the aerodynamic principles, but due to weak materials or engine failures his test planes kept plunging into the Potomac River at the most inopportune times. After his third failure, in 1903, the New York *Times* editorialized that his work was a waste of time and of government funds and predicted that man would not fly for another thousand years. Nine days later the Wright brothers made their successful flight.

Before all these activities in aeronautics, Langley was a fine astronomer. He was an innovator. In 1878 he invented the bolometer, which, as Abbot later described it, was "a wondrously delicate electrical thermometer, capable of measuring to a millionth of a degree, and of estimating the energy of all solar rays in their just proportions."

In 1881 Langley took his bolometer across the country and up the slopes of Mount Whitney, the highest mountain in California. High altitudes reduce the amount of atmospheric absorption to correct for. Langley's determination of the solar constant in that expedition was way too high due to an error in logic he made in calculating the transparency of the atmosphere above the observing site. Had he noticed and corrected that error, his measurement would have yielded almost the proper value of the solar constant. But the measurements of the sunlight did show something else important. In Langley's own words, referring to himself in the third person:

One day in 1881 . . . when being near the summit of Mt. Whitney in the Sierra Nevada, at an altitude of 12,000 feet, he there with this newly invented instrument was working in this invisible spectrum. His previous experience had been that of most scientific men, that very few discoveries come with a surprise; and that they are usually the summation of the patient work of years. . . .

Langley moved his instrument to register farther into the invisible (infrared) portion of the spectrum of sunlight until no more energy was being recorded, "the very end of the end." But he didn't stop there.

By some happy thought he pushed the indications of this delicate instrument into the region still beyond. In the still air of this lofty region, the sunbeams passed unimpeded by the mists of the lower earth, and the curve of heat, which had fallen to nothing began to rise again. These was something there. For he found, suddenly and unexpectedly, a new spectrum of great extent, wholly unknown to science and whose presence was revealed by the new instrument, the bolometer.

Langley had penetrated into a new spectral region of the sun's radiation, the far infrared. The unit of radiation equal to 1 calorie per square centimeter is now called the langley in his honor.

At the time of Langley's Mount Whitney measurements Charles Greeley Abbot was only a boy of eight in New Hampshire. But by 1902, Abbot was very much on the scene. He had joined the Smithsonian in 1895, fresh out of MIT. No new attempts to measure the solar constant had been carried out in the intervening twenty-one years. But the bolometer had been much improved. It was now easier to use accurately than an ordinary mercury thermometer. "Within 15 minutes," Abbot recalled, "we could make a better energy curve of the solar spectrum than Langley and Keeler, with all their skill, could have determined in three days in the year 1881."

The solar constant measurements were now taken up by Abbot at the Smithsonian Institution's Astrophysical Observatory, which was then in Washington, D.C. By chance the few solar-constant determinations made that year and the next seemed to indicate a wild change in the sun's luminosity, a fall of 10 percent. We know now that cannot have happened, and even Abbot thought it likely to be an accidental error of local atmospheric origin. But it stimulated Abbot's interest, and made him a lifelong enthusiast for solar-constant measurements and a promoter of the idea that the sun's luminosity varies.

"Whether true or false," he recalled, "it was the incentive to the long train of observing which has carried our expeditions to four continents, from sea-level to far above the clouds; and at length, after nearly a quarter of a century, it has not only demonstrated the reality of solar changes, but has discovered their unquestionable and important influences upon the earth's temperatures, pressures, and rainfalls."

This was written in 1929, the year after he became head of the Smithsonian. The claim about the reality of the luminosity changes and their terrestrial effects is typical of Abbot's unswerving certainty on the subject. That conviction lasted throughout his scientific career, which would span nearly eight decades!

Abbot was soon off to Mount Wilson, in 1905, for more mountaintop solar-constant measurements, the words of Langley in the last conversation they would ever have (Langley died in 1906) ringing in his ears. "I wish to emphasize that the expedition is primarily to test the variability of the sun. It is not primarily to measure the absolute value of the solar constant of radiation." Actually, this was said humorously, because Langley still hoped his successor would manage to verify his 50-percent-too-high solar-constant determination of twenty-four years earlier. But proof of variable luminosity was Abbot's goal.

Abbot's enthusiasm carried over into his writing about his solar observations. "Those were indeed great days upon Mount Wilson! It was before the completion of the great reflecting telescopes for stellar work. Everybody studied the sun, or some phenomena in the laboratory which were closely related to it. When the sun set, then observing was ended; and in the evenings, with the finest of good companionship, we crowded about the fireplace in the library of the 'Old Monastery.' "

Abbot's book, *The Sun and the Welfare of Man*, published by the Smithsonian in 1929, is a charming mixture of hard science, personal reminiscence, and zesty description.

On a mountain top in Chile [it begins] the Smithsonian Institution maintains a queer observatory. It has no telescope! "Impossible," you say, "an observatory without a telescope." Rather than "on a mountain," it would have been more accurate to have said "in a mountain." The delicate observing instruments are contained in a dark tunnel, over 30 feet deep, running southward from near the summit of the northern face of the peak. The observatory does no work at night, for its studies are confined to a single star, our own star, the sun. This orb is so bright that it needs no telescope to concentrate its rays.

He goes on to describe how sunlight is reflected into the tunnel, where a prism spreads the light into all its colors and instruments measure the heat of each. He describes the barren geography of northern Chile ("neither tree nor shrub, beast nor bird, snake nor insect, not even the hardiest of desert plants is found there"), the weather ("hardly ever does rain fall"), and the lonely life the observers have "for three years at a shift . . . dwelling in this wilderness."

The reason for all this sacrifice? It is almost his personal anthem: "For the purpose of measuring that solar energy which supports every form of life and activity upon our earth, and especially of noting its changes. For if our supply of heat from the sun alters, weather and crops must be affected. If the sun's output of energy should permanently diminish or increase by considerable amounts, the whole future of civilization would be destroyed."

Abbot continued to carry out solar-constant observations all his life. In the late 1920s, with National Geographic Society assistance, he established still another observing site on the top of Mount Brukkaros in South-West Africa, "the best site for a solar radiation observatory in the Old World." Still other sites were later set up in North Africa and the Middle East. The goal of all this enterprise was to gain satisfactory observations of the sun on every day of the year by at least two stations, have no more than one third of 1 percent discrepancy in the readings between stations, and to maintain such observations continually for twenty years. The daily observations were, in fact, carried out longer than that, until 1952.

During and after his years as secretary of the Smithsonian, Abbot continued his solar-constant work from Washington. He retired as secretary in 1944 at the age of seventy-one, but in succeeding decades still maintained his office in the Smithsonian castle headquarters on the Mall and kept up his interest in the variable solar constant. He never wavered from his belief that the sun's luminosity fluctuated. As he put it, "The Astrophysical Observatory has indeed discovered that the sun is a variable star, and for many years has observed its variations. . . . The variations . . . are small in percentage and apparently irregular in amount."

Abbot was a Washington institution. In one sense he was ahead of his time. He invented many practical devices for harnessing solar energy, including the Abbot solar cooker, which he used to cook meat and bake bread during the long stays at Mount Wilson. He was respected for his enthusiasm and perseverance. I remember seeing him at a meeting in Washington honoring him on one of his "upper nineties" birthdays about 1970. Very tall and slender with white hair and white mustache, he walked with a cane but still seemed erect and alert. Unfortunately, for all the respect they accorded him, few scientists shared his view that variations in the sun's luminosity were the driving influence on weather.

Charles Greeley Abbot died December 17, 1973, at the age of 101.

Abbot's long record of solar constant observations suffered from two unavoidable circumstances. They necessarily had to be made within earth's atmosphere, and the measurement technology was not accurate enough to be unmistakably persuasive. In 1976, at a conference on the solar output and its variation, scientists offered cautious evaluations of the nearly fifty-year record of daily Smithsonian Astrophysical Observatory measurements of solar luminosity under Abbot's direction. The program had reported changes in luminosity of from 0.1 percent to 1 percent and also suggested an apparent long-term increase of the average value of the sun's luminosity by about 0.25 percent in fifty years.

"A conservative summary of the Smithsonian measurements," said Claus Fröhlich, a solar-constant specialist from the World Radiation Center's Physico-Meteorological Observatory in Switzerland, "is that changes of more than 1 percent in solar luminosity were not seen and changes of less than 1 percent were not capable of being seen."

John Eddy echoed this pessimistic appraisal. "Current assessment of the SAO data," concluded Eddy, "suggests that most or all of the changes reported are probably indistinguishable from uncertainties in atmospheric transmission, for which corrections were necessarily large, and from changes in instruments, calibration, and technique." In other words, there could be no certainty that the reported changes of the sun's output in Abbot's painstakingly compiled record were real. The scientists were still on square one.

Nineteen seventy-six, you will recall, was the same year the scientific community had been jolted into a new realization that the sun's behavior had not been regular in the past. Eddy's report about the Maunder sunspot minimum had just come out, and Eddy had by now already extended the record of solar irregularity back to the Bronze Age.

In addition there was a new awareness throughout the world

of the sensitivity of agriculture and the world's economies to small changes in climate. The Sahel drought in Central Africa was on everyone's mind, and unexpectedly bad weather had damaged crops in 1972, 1974, and 1975. Climatologists were warning the public that the great rise of American agriculture in the first half of the twentieth century had taken place during what they now realized had been an unusually warm and mild period of climate. The newly revealed long-term record of climate indicated that more variable weather was typical. In future decades we should expect a wider swing in the year-to-year fluctuations of weather. Possible variations in the sun were being considered as one possible factor in climate change. The whole subject of solar variability had gained new urgency and respectability.

Yet here we were still with no reliable record to show whether the sun's luminosity—the solar constant—changed. Scientists suddenly realized the opportunity they had missed. We had been launching satellites and spacecraft for the better part of two decades. With only a couple of exceptions, no one had put solar-constant measuring instruments on board any of them. There, above all the earth's absorbing atmosphere, the needed measurements could presumably be made with the precision (about 0.1 percent) never before possible. So no long-term record of solar luminosity had been compiled from space. As for the measurements that had been made, the differences between them often exceeded their accuracy. Their precision was doubtful.

"In my opinion," said Stephen H. Schneider, a young climatologist concerned about all these matters, "this measurement of solar variability could turn out to be the most important single geophysical observation that could be made (from space or earth). The lamentable oversight that permitted the world to watch men skipping, planting flags, playing golf, saluting presidents, and race driving across the moon without quietly placing an instrument capable of shedding light on the debate as to whether variations in the sun cause variations in the climate seems one of the major scientific omissions of the 1960s!"

Schneider called the measurement "fiendishly difficult and costly, but I think justifiable." He compared the lack of it to the problem a family or business has trying to prepare a budget without knowing the external influences that could affect it.

Eddy too regretted the lack. "It is unfortunate that the continuous measurement of the solar constant has never been given priority in science, either by funding agencies or by atmospheric or solar physicists." The reasons were simple enough. "It is a difficult measurement; it is an unexciting measurement; it requires dedication and funding over periods not in years but in decades. Modern science does not seem set up to tackle problems of this time scale."

Scientists generally aren't the type to lie down in the face of a challenge, however. Until the necessary long-term records can be compiled from space, beginning in the 1980s, they have been searching for other ways to extract meaningful information on the sun's luminosity. The past five years have brought many such efforts. In fact, we've seen a regular outburst of new attempts to measure the solar constant, all directed to learning if our sun is indeed a variable star. But the real question is not just whether the sun's output of light varies, but by how much

In January 1976, Richard C. Willson, a physicist and meteorologist at the Jet Propulsion Laboratory in California, launched an Aerobee rocket containing four independent solar constant instruments. Comparing the results to those from a high-altitude balloon flight he had launched in 1969, Willson concluded that the solar "constant" had remained unchanged to within 0.75 percent over those six years, the last half of solar cycle 20.

Then in 1978, University of Denver physicists led by David G. Murcray launched a high-altitude balloon carrying an instrument to measure solar luminosity. Murcray's group had flown the same instrument a number of times in 1967 and 1968 and then, as he says, "put it on the shelf." With all the new interest in climate change and the sun, they decided to fly it again. It was carefully recalibrated to make sure its accuracy had remained consistent. The instrument was launched January 27, 1978, and its readings then carefully analyzed. Murcray reported the results at a meeting in Washington in 1979.

Solar irradiance as measured by the balloon-borne instrument had increased by 0.4 percent between 1968 and 1978. "This change is greater than the experimental uncertainty and is felt to be the result of change in the solar constant," reported Murcray. "We think it is a solar effect. The most logical point here is that these data represent a change in solar output."

He pointed out that the increase of solar radiation by 4 parts in 1000 was smaller than previous experiments could have detected.

In early 1980 Willson was back with new rocket measurements. The flight had been made in November 1978 under conditions almost identical to those of the 1976 measurement. The instruments had been aimed directly at the sun during the five minutes they had spent between 100 and 250 kilometers above the earth, but the key observations were those near the peak of the trajectory. Willson and colleagues C. H. Duncan of NASA and J. Geist of the National Bureau of Standards announced the results in *Science*: The total solar output of radiation, correct to the average sun-earth distance, was 0.4 percent higher in 1978 than thirty months earlier in 1976. "There is a high probability that this measured difference is real," they said. In fact, they felt their experimental result was the first measurement of solar luminosity from space in which the uncertainties in long-term precision were no more than 0.1 percent.

This result caused considerable discussion and excitement. The amount of change indicated in the sun's luminosity in just two and a half years was surprising. Did it represent some random fluctuation in the sun, an isolated change lasting only a short time, or a sustained change important to earth's climate? There was no way of knowing.

Unfortunately, as often happens in science, the result did not long hold up. In the fall of 1980 Willson and his colleagues did further tests on their instruments and data that allowed them to evaluate better the rocket experiments. The tests also allowed them to narrow the error in measurements they had made from a balloon back in 1969. During the 1978 rocket flight, the solar luminosity instrument had been subjected to slightly elevated air pressures. That condition, the new tests showed, caused it to give an unjustifiably high reading. The corrected value no longer showed a 0.4 percent incease in solar luminosity between 1976 and 1978. There was no increase at all. If anything, the corrected 1978 value was slightly lower. The same correction also had to be applied to results of another rocket flight experiment flown in May of 1980.

The corrections placed tighter constraints on the amount of change observed in solar luminosity from year to year. Said Willson in January 1981, "Over the 11 years from 1969 to 1980, we now know the sun hasn't changed by more than 0.2 percent, and probably the change is less than that. And between 1976 and 1980, we feel confident that the change was less than 0.1 percent."

Willson's readings were thus indicating less than half as much change in the sun's output between the late 1960s and the late 1970s as were Murcray's. Willson's experiments had smaller experimental uncertainties, however. That gave his results somewhat more credence.

All these measurements were taken at isolated and widely separated times, so they are relevant only to changes in the sun's luminosity over fairly long periods of time. They said nothing about possible short-term fluctuations.

Some insight into how quickly solar luminosity apparently could change came from the Nimbus 7 meteorological satellite. Nimbus 7, launched in November 1978, carried a self-calibrating solar-constant sensor. Lead scientist J. R. Hickey reported in 1980 that it had detected some remarkable changes. Nimbus 7 had started its observations during the period of increasing solar activity for solar cycle 21. August 1979 was a month of high solar activity, and Nimbus 7 found a temporary *decrease* in solar luminosity of 0.36 percent. In early November 1979 the sun unleashed a series of very strong flares. Again, Nimbus 7 measured a drop in the sun's luminosity. The solar constant fell up to 0.4 percent during those events. This was one of the first reliable indications that the sun's luminosity can change quite quickly as a result of events associated with solar activity.

Rapid changes in certain wavelengths of the sun's radiation—the ultraviolet, X-ray, and radio portions of the spectrum—are expected at times of high solar activity. That's because solar flares emanate strongly in their invisible wavelengths. But 99 percent of the sun's luminosity is in visible light, and even dramatic bursts of radiation in these other invisible wavelengths shouldn't add to the sun's general luminosity by enough to be measured in solar-constant observations. The amount added should be only about 0.01 percent. Besides, Nimbus detected a *decrease* of luminosity in these highly active periods, not an *increase.* That is a puzzle.

Solar scientists have long been intrigued by a possible relation between sunspots and the solar constant. Sunspots are obviously darker than the rest of the sun, so at times when sunspots are prevalent shouldn't the total amount of sunlight emanated be reduced? It's an old idea. Such an effect might be expected to reduce the solar constant by several hundredths of a percent. But it's not all that simple. The energy blocked off by spots might well emerge instead as a slight increase in radiation somewhere in the 99.9 percent of the sun not covered by sunspots. That would cancel the sunspot effect. And bright regions on the sun called faculae tend to be larger when sunspots are abundant. So they too could cancel out the darkening effect of sunspots, even to the point of increasing the solar constant.

Recently Peter Foukal, a scientist interested in the solar-constant question, and colleague Jorge Vernazza made a statistical analysis of thirty years of the Smithsonian Astrophysical Observatory's daily solar-constant data. They wanted to see if any effect of the sun's rotation showed up. If there were some fluctuations that repeated at intervals of twenty-seven days—the rotation period of the sun—that would be an indication the changes might be due to large sunspot groups rotating with the sun. They found luminosity variations of about 0.07 percent tending to recur at the period of the sun's rotation. They also found that the solar constant drops with increasing sunspot area and rises with increasing area of the bright faculae.

This could all be explained if indeed the magnetic regions associated with visible sunspots do temporarily obstruct, redirect, and modify the flow of heat to the solar surface.

That's the way things stood until the new era of space-based monitoring of the solar constant began in 1980. On February 14 of that year NASA launched the Solar Maximum Mission spacecraft into a circular orbit 660 kilometers (410 miles) above the earth. There, a complex set of instruments focused on solar flares in an effort to catch them in early stages of birth and observe them throughout their short lifetimes.

The SMM spacecraft also carried a set of sensors capable of measuring the sun's total luminosity with a precision and accuracy never before achieved in satellite observations. Called the Active Cavity Radiometer Irradiance Monitor experiment, it was designed at the Jet Propulsion Laboratory by Richard Willson and his colleagues. It was the first of the new generation of space-borne devices designed to compile the precise long-term record of variability in the sun's luminosity that scientists had begun calling for four years before. The instrument could detect variations as small as 0.001 percent, or 1 part in 100,000.

By September, Willson had analyzed the first five months of

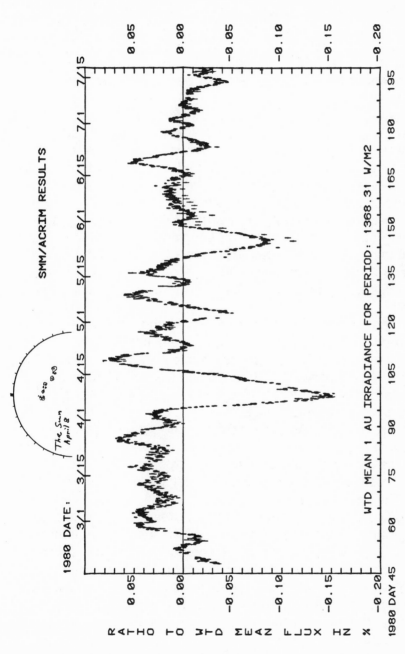

The sun's total irradiance over the first 153 days of the Solar Maximum Mission in 1980. There is a continuous variation of the total solar flux below the 0.05 percent level and two large, temporary decreases of 0.1 to 0.2 percent centered on April 8 and May 25, each lasting more than a week. Both these dips came when large sunspot groups were crossing the central solar meridian. A drawing of the active regions on the northern half of the sun's disk on April 8 is superimposed at the top. (Courtesy Richard C. Willson and Jet Propulsion Laboratory)

the solar-constant data. The conclusion was unmistakable. The solar constant definitely varies. The satellite showed the sun's luminosity fluctuating almost day to day. The amounts of change were generally small, about 0.05 percent, or 1 part in 2,000. But in those first 153 days of observation there were two much larger dips in luminosity, representing changes of up to 0.2 percent, or 1 part in 500, over a period of seven to ten days.

Willson called the continuous variability of the luminosity and the two large temporary decreases "striking features." Jack Eddy called the Willson results "exciting," the most important new finding about the sun in the past year. As Eddy told me: "They are detecting changes in the solar constant that are believable. They are small, but real. It is the first time that has been confirmed."

The long quest to prove a variable solar constant that had begun with Pouillet in the 1830s and Langley in the 1880s and had continued through the seven-decade career of Charles Greeley Abbot into the rocket and balloon observations of the late 1970s had finally, belatedly, entered the space age. The answer was certain. The sun is a variable star.

What was perhaps most remarkable about the observations was that the fluctuations in solar irradiance were highly correlated with a newly devised index of the area on the sun covered by sunspots. The strong dips in luminosity around April 8 and May 25, 1980, came just as large groups of sunspots were crossing the center of the sun's disk. The relationships between these two sunspot groups and the sunlight variations were obvious. As each of the groups approached the center of the central meridian of the sun, solar luminosity nosedived. The depth of the dip came just as the spots crossed the sun's center. Luminosity rose rapidly again as the spots left the central meridian.

The largest of the two drops in luminosity came on April 8. At that time 80 percent of the indexed sunspot area was due to two large sunspot groups near the center of the sun. The luminosity dip of May 25 was not quite so low. On that day 61 percent of the sun's sunspots were near the center of the disk. In other words, the spacecraft observations were showing not only that the main luminosity variations were due to passage of sunspots across the center of the sun but also that the degree of the luminosity decline depended upon the size of those sunspots.

Apart from these two main plunges in solar output, the

correlations between luminosity and the sunspot-area index were generally good. Some differences were apparent. There was an intriguing five-day drop in solar irradiance centered on May 3, 1980, that seemed to have no counterpart in visible sunspot activity. That fact, as Willson noted, supports the probable existence of other sources of solar variability. The list of such sources, as I said in chapter 2, is large, and new ones are continually being discovered.

All in all, the association of large sunspot groups with declines in luminosity supports the view that sunspots can temporarily impede or divert the upward flow of energy in the upper part of the sun. As Willson and coworkers put in the paper they prepared in September 1980 for *Science*: "We are led to the tentative conclusion that the sunspot-deficit energy is stored or delayed within the convection zone through the effects of the magnetic fields associated with the sunspot groups involved."

Exactly what happens to that stored energy wasn't clear. There was no sign that it was quickly released to the bright faculae. If it were, the correlations seen would have been much weaker. If it does emerge in these bright regions, it must do so only after some considerable delay. Perhaps faculae that develop later in these regions radiate the stored energy away. Peter Foukal had earlier suggested that the powerful coiled magnetic fields below the solar surface might store energy so well that it wouldn't have to emerge through bright faculae and flares until perhaps years later. The answer to all that would have to await further study.

For now, Willson's Solar Maximum Mission luminosity-monitoring instrument had done enough. In just five months of operation, it had shown that the solar constant fluctuates almost daily, that large dips in luminosity are caused by passage of major sunspot groups across the center of the sun's disk, and that considerable quantities of energy must be temporarily stored beneath the surface of the sun in magnetically active regions.

It had demonstrated beyond doubt that, in changing its output of light by several tenths of a percent in periods as short as a week, the sun is indeed a variable star. Given the century-long debate over that very question, who could ask for more?

Richard Willson wants to carry on these measurements until the year 2000. The space shuttle offers the chance to do that successfully. It will carry much-improved solar-luminosity instruments into earth orbit in the 1980s. There they can be left for whatever length of time is desired. Any necessary reservicing can be done in space. And when an experimental run is completed, the shuttle can return the instruments to earth so the scientists can carefully recheck their accuracy in the laboratory. Then they can be returned to space for more observations.

The satellite work has already shown conclusively that the sun's output of light does fluctuate. Two continuous decades of such precise measurements will provide a richly detailed record of solar-luminosity variability, an invaluable legacy to future generations of inquirers. Then we'll better understand our sun's nervous habits. And we should have the key to answering to what degree the sun's variable luminosity can account for any climate fluctuations, past or future. Charles Greeley Abbot's long dream will at last have been fulfilled.

8. QUEST FOR THE CLIMATE CONNECTION

Much has been written [on the influence of solar activity upon the lower atmosphere, in terms of weather and climate] and many claims have been made, but most have been refuted by later work.

Astronomer M. A. Ellison, *The Sun and Its Influence*

In the 200-year history of sun-weather studies, a large body of information has been accumulated. Even though the reported results have sometimes been confused, disjointed, and contradictory, there has emerged a growing belief that there are connections between changes on the sun and changes in the lower atmosphere.

John R. Herman and Richard A. Goldberg, *Sun, Weather and Climate*

Although the literature on [solar-weather relationships] is voluminous and extends back in time for well over 100 years, the field remains in limbo. There exists no unequivocal empirical evidence of a connection between solar energy variations and weather.

Geophysicist Ralph Shapiro

The effects that we observe of solar activity on the weather are, one

should admit, small and even esoteric. But they do apparently cause
discernible responses in the weather.

Solar physicist Walter Orr Roberts

The history of inquiry into effects of a varying sun on weather and climate is long and inglorious. Few areas of science have been pursued by so many for so long a time with so little of certainty to show for the effort. Like the course of an overimpetuous love affair, there have been bursts of passionate enthusiasm followed by sudden disillusionment, only to be followed by hopeful new attempts to establish the sought-for connection.

The sun has been a tantalizing and coy seductress. Its variable face has tempted generations of scientists to try to associate its cycles with patterns of events on earth. More often than not, short-term success has been followed by long-term failure.

That's not altogether surprising. Two cyclic patterns may, by coincidence, appear to march in step for a short time; a longer view often shows no real association. If we have only the short view, we may be misled.

Also, the search for correlations of any sort is strongly influenced by psychological matters, such as our own hopes and expectations. This can result in an unconscious selection process. We tend to notice those connections that we seek and ignore all those items that don't fit. The result sometimes is to assume connections when there is no real basis for doing so. We all do that in our daily lives. ("I was just thinking of Uncle Ernie, and he called! Isn't that amazing!") Scientists are as human as the rest of us. They can be susceptible too. The self-correcting process of science usually eventually roots out and rectifies the biasing influences of human perception and judgment. But that often takes some time.

Quests for practical terrestrial effects of a variable sun go back at least to 1801. That's when the eminent English astronomer Sir William Herschel suggested a possible correlation between the amount of sunspots on the sun and the market price of wheat. If you look at a chart of sunspots and wheat prices during the three or four decades before Herschel's proposal, you can see what he was talking about. Both do fluctuate upward with about ten-year intervals (although the rises seem out of phase to me). But in the decades immediately after 1801 the two cycles no longer appear to have any resemblance to each other. Herschel's necessarily short-term view had been deceptive.

By the 1870s and 1880s, the fact that sunspots apparently do come and go in more or less regular cycles was now common scientific and popular knowledge. The clear associations between the sunspot cycle and auroras and magnetic storms on earth had been noticed. It wasn't long before sunspots were being associated with all manner of other terrestrial events.

Any cyclic occurrence seemed fair game for those who proposed links to sunspots: the periodic visitations of Asiatic cholera, the ups and downs of the world's financial markets, temperatures in tropical rainy regions, wind directions in Argentina, the rabbit population of England, the levels of Lake Victoria, the depths of the Nile and the Thames, monsoons in India, and soil temperatures in Scotland.

Some of the proposed connections were not wholly arbitrary. Financial crises could be brought on by crop failures, which are caused by bad weather, which might be stimulated by variable solar heating or magnetism. Other suggested links were only statistical associations with no rationale supporting them.

The scientific community, then much as now, was divided on the sunspot-weather question. Sir Norman Lockyer, the eminent English solar astronomer* who founded and edited for half a century the respected British journal *Nature*, considered the influence of sunspots upon storms and rainfall to have been conclusively demonstrated. Lord Kelvin, in contrast, didn't even admit the reality of the connection between sunspot cycles and magnetic storms on earth, which most scientists by then readily (and correctly, it turned out) accepted. In his presidential address to the Royal Society in 1892 Kelvin called that proposed connection "unreal" and "a mere coincidence." But by then Kelvin, following a brilliant scientific career, had become something of a scoffer at all sorts of new ideas sweeping the scientific world. (In his eighties he bitterly opposed the idea that radioactive atoms were disintegrating.)

The American solar astronomer Charles A. Young aptly summarized the situation in his book *The Sun*, published in 1895:

Lockyer discovered helium on the sun, by examining the spectral lines of sunlight, in 1868, nearly thirty years before it was identified in the laboratory! He named it helium from the Greek word for the sun, helios.

In regard to this question [whether sunspot periodicity pro-duces notable effects upon earth] the astonomical world is divided into two almost hostile camps, so decided is the difference of opinion, and so sharp the discussion. One party holds that the state of the sun's surface is a determining factor in our terrestrial meteorology, making itself felt in our temperature, barometric pressure, rainfall, cyclones, crops, and even our financial condition, and that, therefore, the most careful watch should be kept upon the sun for economic as well as scientific reasons.

The other party contends that there is, and can be, no sensi-ble influence upon the earth produced by such slight variations in the solar light and heat, though, of course, they all admit the connection between sunspots and the condition of the earth's magnetic elements. It seems pretty clear that we are not in a position yet to decide the question either way; it will take a much longer period of observation. . . . At any rate, from the data now in our possession, men of great ability and laborious industry draw opposite conclusions.

It is ironic that almost the same words could be applied to the subject today. The sunspot-weather connection has been one of the most troublesome and elusive concepts scientists have faced in the past century.

The problem was that the simple and seemingly certain associations people thought they saw in the 1870s and 1880s didn't hold up when the longer-term record began to be compiled in the twentieth century. Many of the connections that did continue in the records were small and subject to personal interpretation.

Most scientists learned to be cautious in asserting a sun-weather connection. But sunspots had entered the folklore. Through-out a good portion of this century sunspots continued to be half-seriously invoked to explain all manner of things in much the way you often heard changes in the weather in the 1950s and 1960s blamed on "those atomic bomb tests."

In *Other Worlds*, an informal little book of personal observa-tions, the American astronomer Carl Sagan relates an anecdote about a talk he and the Soviet astrophysicist I. S. Shklovskii were listening to at a scientific meeting. The speaker seriously suggested that the world's greatest scientific achievements had been accomplished at times of maximum sunspot activity. The great works of Newton,

Darwin, and Einstein had, so said the speaker, been done during solar maximums. Shklovskii, who has a well-tuned sense of humor, leaned over to Sagan and whispered loudly, "Yes, but *this* theory was conceived in a deep solar *minimum*."

Another story about the pitfalls of an overeager search for correlations was related by the late Harvard astronomer Donald Menzel in his book *Our Sun*. A British scientist interested in possible relationships between the weather and times of birth checked the records and found to his surprise that babies tended to be born more often on clear days. Another scientist who read the report in a scientific journal checked the mathematics and found nothing wrong. Still skeptical, he went to the agency that had supplied the birth information. Everything was correct, except for one thing. The clerk had mistakenly entered not the date of birth but the date the parents had registered the birth. All the data showed was that people are less likely to go out in bad weather!

In the late 1970s, several books appeared proposing strong correlations between terrestrial events and phases of the moon. One popular idea was that more babies are born when the moon is full. Two scientists at the University of California at Los Angeles, astronomer George Abell and physician Bennett Greenspan, found that even the obstetrical nurses at the UCLA Hospital had no doubt that it was true. It seemed to them that more babies *were* born during the full moon. Abell and Greenspan checked the statistics of all eleven thousand births at the hospital during the previous four years. There was absolutely no correlation with any phase of the moon! Apparently the nurses simply tended to remember those months when there was a full moon during a particularly busy night, and they forgot the rest. The perils of human subjectivity!

Another spurious solar system/earth link, this one specifically invoking the sunspot cycle, got widespread popular attention in the 1970s. Two astrophysicists wrote a book called *The Jupiter Effect*. They proposed a complicated series of geophysical links in which a lineup of the planets in 1982 would coincide with the peak of sunspot activity. This would cause especially high solar activity, which would change air patterns in earth's atmosphere, which would alter the rotation of the earth, which would cause an incredible series of giant earthquakes in California.

Each of these connections is scientifically flimsy, and strung together they amount to flimsy compounded. Besides, the planets wouldn't be lined up in 1982 or in any other year; they simply would all be on the same side of the sun. That has happened many times before with no apparent ill effects.

No scientists took the Jupiter effect idea seriously. But many nonscientists did. Cult groups took up the cause, and by 1980 there was some mild public hysteria about the coming sunspot-related unprecedented earthquake tragedy in California. (A giant earthquake there may be; scientists have long warned one is overdue. But that has nothing to do with solar activity.) The situation got so bad that in the summer of 1980, the lead author of *The Jupiter Effect*, John Gribbin, wrote a public retraction. "I have bad news for doomsayers," said Gribbin. "The book has now been proved wrong; the whole basis of the 1982 prediction is gone." The sunspot cycle, it turned out, had already peaked in late 1979 and early 1980, two years before the planetary (non)alignment. So the first link in the proposed chain of connections had been decisively severed. And none of the other links had gained any support either. "If anyone tries to warn you about the Apocalypse coming in 1982," Gribbin said forthrightly, "just tell him that the old theory has long since been disproved."

The late American Nobel laureate scientist Irving Langmuir coined the term "pathological science" for what he called "the science of things that are not so." It was a reference to the natural human tendency by scientists occasionally to become strongly committed to certain beliefs in the face of evidence to the contrary. It is more even than that. Langmuir referred to the way scientists often see what they want to see in "noisy" data. He cited examples of chemists seeing responses that their more dubious colleagues, standing beside them, could not see.

Only three years ago, one of the harshest critics of proposed sun-weather connections, an Australian atmospheric physicist named A. Barrie Pittock, invoked an earlier warning of a Russian scientist, Andrei S. Monin. Monin said those scientists who see evidence for the influence of solar activity on the weather are engaged in "successful experiments in autosuggestion." In other words, suggested Pittock, they were deluding themselves. The connections were in their minds, not in the data. And he was talking about responsible astronomers and meteorologists, engaged in research on sun-weather effects, not popular writers and mystics.

That is a harsh appraisal. Is it justified? Certainly, there have been instances where it is true. But it is an extreme position. Most workers take a more moderate view. It does serve a gadfly purpose. It reminds scientists seeking the sun-weather connection of the perils of the past and the pitfalls to avoid in the future. Most of these scientists are well aware of them. They have to be. They have seen the tortuous path the sun-weather idea has followed during the last hundred years, every gain seemingly followed by a setback. The patterns and connections, if they are there, have steadfastly resisted the efforts of bright and exceedingly persistent scientists to unveil them. They are cloaked by all manner of complications. These difficulties begin with our lack of a full understanding of the sun. They get worse when we encounter unpredictable vagaries of our amazingly complex atmosphere. Scientists engaged in sun-weather research have learned to be humble.

I have so far intentionally emphasized the cautious view. I think it is healthy to approach any speculative subject with a realization of past and (probably) future hazards to overoptimism. The scientific adventure is a combination of bold thrusts into the mysteries of the unknown and careful and rigorous examinations into the meaning (if any) of what the explorers find. With these warnings buzzing in our ears, let's now plunge headlong forward into the quest for the sun-climate connection.

Well, not forward. Not quite yet. Let's go back a bit first.

Charles Greeley Abbot, who was with the Smithsonian Institution continuously from 1895 until his death in 1973, was certain not only that the sun's luminosity varied but also that these variations influenced weather and climate. "The changes [in the sun] produce effects on atmospheric temperatures and pressures, and on precipitation," he said. "Whether they and their effects shall become predictable for seasons or years in advance, we cannot as yet foretell.... But the work must go on for years before a long enough background will be laid to justify hope of seasonal forecasting."

The year before Abbot joined the Smithsonian, a twenty-seven-year-old astronomer named Andrew Ellicott Douglass made a big decision. He left the Harvard College Observatory and moved out to the sparsely populated clear-sky land of the Southwest to help Percival Lowell found a new observatory on the flanks of the mountains outside of Flagstaff, Arizona. There Douglass, as a hobby, took up the study of tree rings. It was clear that the width of the annual growth rings varied from year to year, and that these variations reflected fluctuations in local weather conditions. Douglass became interested in the possibility that tree rings might be able to reveal something about the effect of the sun on climate.

In the process he became the acknowledged founder of the modern science of dendrochronology, or tree-ring dating, which has proved indispensable to archaeology and climatology. The major work was done at the University of Arizona in Tucson, where Douglass had moved in 1906 to join the astronomy faculty. In 1937, at the age of seventy, Douglass founded the university's Laboratory of Tree-Ring Research, the world's leading center for such studies. Pioneering the science of dendrochronology is what Douglass is primarily remembered for.*

But his original interest with trees was the sun. If the eleven-year sunspot cycle influenced the weather on earth, he reasoned, then some evidence of an eleven-year cycle should show up in the variations of tree-ring widths. He worked carefully on the problem, identifying all the many complications that affect tree growth. He invented optical instruments for identifying recurring width patterns.

Douglass became convinced that there was an eleven-year cycle in the variations of tree-ring widths. It was subtle and not always easy to see. It certainly didn't appear in every tree. Whole regions failed to show it at all. One of the clearest places it showed up was in a controlled forest of Scotch pines grown over an eighty-year period in Germany outside of Berlin. Douglass argued that it was ideal for exhibiting a solar effect because interfering influences such as insect infestations and overcrowding were less of a factor.

Douglass died in Tucson in 1962 at the age of ninety-four, still convinced that the solar cycle showed up in tree-ring widths. Other scientists were not so sure. Skeptical investigators could argue that the seen correlations between width patterns and the sunspot cycle were accidental and scattered, untypical either of the long-term record or of most trees in general. Proponents could point to clear correlations in at least isolated parts of the record for certain regions.

Dendrochronology is a sophisticated science. Saying its practitioners merely "count tree rings" is like saying astronomers just "look through telescopes."

It was a matter of judgment. The case remained unsettled.

Nearly half a century after Douglass first went west to help set up Lowell Observatory, another young astronomer traveled west from Harvard. His name was Walter Orr Roberts, and he was destined to become a leading figure on the subject of sun-weather relations.

Roberts had come to Harvard as a fresh young graduate student in 1938. Unable to take all the physics courses he wanted because of a conflict, he enrolled in an astronomy course taught by Donald Menzel. Menzel was a stimulating teacher. It took Roberts only a few months to decide to switch to astronomy and to join up with Menzel in an exciting new project. Menzel had designed a new astronomical instrument, a solar coronagraph. It would enable them to observe the tenuous outer atmosphere of the sun, called the corona, extending far out into space. Normally the corona is invisible, due to the glare of the sun, but the new instrument would block out the glare and allow artificial eclipse photography of the corona.

Menzel had grown up in Colorado, and lived as a boy in Leadville, at 10,200 feet the highest community of any size in the United States. He decided to build the new coronagraph and locate it in the Colorado mountains to gain the advantage of clear and thin air. He chose a site on Fremont Pass, at the mine of the Climax Molybdenum Company, altitude 11,520 feet. He persuaded Roberts to move west to operate the observatory.

"There I went in July, 1940, with my bride of about three weeks," recalls Roberts, "in an old broken down Graham Paige car, with most of the coronagraph in it."

The car immediately broke down in Wellesley, Massachusetts, but he and his wife Janet eventually made it to Climax. There they found that the beautiful house Menzel had arranged for them to live in was built exactly on the continental divide.

"The water that drained down the east side of the roof went through my rhubarb plants and down into the Atlantic Ocean. On the other side, it went through my wildflower garden and down into the Pacific Ocean." It didn't take Roberts long to make up a joke about how during the war years, he and Janet would check the newspapers to see which part of the country was experiencing shortages of water and throw their dishwater out the appropriate side of the yard.

During those war years, Roberts quickly measured the daily photographic plates of the corona taken at Climax "even before they were washed or were dried" and transmitted relevant information to laboratories in Washington responsible for radio propagation forecasting. He and another scientist had independently discovered a relationship between the corona and the quality of radio communications, a matter of crucial importance in a war.

Some of the earliest financial support for the Menzel coronagraph had come from the U.S. Department of Agriculture under Secretary Henry Wallace. The mid-1930s were the Dust Bowl years, and partly due to Charles Greeley Abbot and others officials were interested in whether the comings and goings of great droughts were related to solar-activity cycles. The evidence was not yet in.

Roberts became interested in the idea in the early 1950s. By that time he had moved off the continental divide down to the base of the Rockies at Boulder, where he, Menzel, and another Harvard astronomer set up a center called the High Altitude Observatory, which Roberts directed until 1960. HAO was for many years affiliated with both Harvard and the University of Colorado. It is now part of the National Center for Atmospheric Research in Boulder, which Roberts headed, first as director and then as chairman of the parent University Corporation for Atmospheric Research, from 1960 to 1973.

The Great Plains droughts and alternate sunspot cycles did both seem to come at twenty-two-year intervals. The droughts brought awful human tragedy. Water is *the* critical influence on the economies of the West. Any edge gained in better anticipating and understanding the periodic onset of droughts would be a godsend.

Twenty-two years is a long time. Many such cycles would need to be studied to establish any good statistics on the matter. "I was convinced, however, that if there is indeed a sunspot-cycle relationship, there must also be relationships in the short term," Roberts recalls.

Roberts had hired a young colleague to do a survey of the relevant scientific literature. Using Air Force weather data he found differences in the patterns of barometric pressure in the northern hemisphere in the days after geomagnetic storms compared with days of geomagnetic quiet. "It suggested," Roberts recalls, "that there was some barometric change due to solar activity."

Roberts encouraged another colleague of his, Ralph Shapiro of the Air Force Cambridge Research Laboratories near Boston, to test the proposition. Times of large perturbations in earth's magnetic field were then still the best proxy indicator of solar activity known to affect earth. Shapiro studied the weather maps and geomagnetic records for the previous half century.

Shapiro's work showed a distinct pattern in the average persistence of weather patterns of North America in relation to times of geomagnetic storms. By four days after geomagnetic storms the persistence is significantly above normal, meaning the pressure patterns are changing only slowly. By fourteen days after magnetic storms persistence drops way below normal, meaning pressure patterns were changing rapidly with time.

Was it real? "This result, we believed then and still do, was significant statistically," says Roberts. "If you divide the data set in different ways you still find the same results."

That, says Roberts, is what really stimulated his deep interest in the sun-weather connection. He and his colleagues in Boulder pored over weather maps. The apparent geomagnetic effect on weather seemed to be most pronounced on storm systems in the Gulf of Alaska. The gulf is a large section of the North Pacific where many of the storms that affect North America get their start. What they saw was an apparent amplification of low-pressure storm troughs during the days following geomagnetic storms. Low-pressure troughs that entered the gulf two to four days after geomagnetic activity increased in size dramatically over the next four days. Troughs entering the gulf at magnetically quiet times showed no such growth.

Once again they had statistics to test the findings' significance. The probability of the distribution they were seeing occurring by chance anytime in a three-year period was something like one in a million. "We were highly confident of our results," says Roberts, "but I didn't have much luck selling it to skeptical colleagues." The reason? There was no satisfactory physical explanation of what was happening. That is still true.

Nevertheless, Roberts was persuaded that something important had been discovered. This sun-weather effect seemed real, and it even appeared to have a potential impact on weather forecasts.

In the 1960s and 1970s Roberts pursued many other scientific and administrative pursuits. But he maintained his strong interest in sun-weather effects. I'll take up the more recent continuations of this work by him and colleagues in later chapters.

Roberts has always taken a broad view of the scientific

enterprise. A friendly, active man with combed-back hair and a ready smile, he has for many years been an able scientific statesman, interested in many questions concerning public issues and science. He's what you might call an "overview" man. He has been president of the American Association for the Advancement of Science and is a fellow of the American Academy of Arts and Sciences and a member of the American Philosophical Society. He has written often and well about the probable effects of future weather and climate change and the need for efforts to better prepare for them. ("There are very few factors in the socio-economic environment that exceed weather and climate in importance to the production of food, to the conduct of commerce, and to the pursuit of a better life.") And he has stimulated a generation or so of scientists also to concern themselves with such matters. He is now director of the Program in Science, Technology and Humanism of the Aspen Institute for Humanistic Studies (whose offices are not nestled beneath the slopes of Aspen but on the second floor of an old-fashioned wooden building near the University of Colorado campus in Boulder).

Walter Orr Roberts remains a solid proponent of sun-weather effects, but a cautious and philosophical one. He recognizes that the search has been filled with disappointments, frustrations, and over-optimism. Charles Greeley Abbot, he says, "went overboard on the subject." He urges all to listen to Andrei Monin's warning about "heliogeophysics enthusiasts." None of us, he says, wants to spend time "following false scents." Satisfactory identification and documentation of sun-weather connections is a slow process. It requires patience. Nevertheless, he considers the evidence that certain weather and climate phenomena are linked with solar activity "impressive."

The possible benefits are enormous. He believes study of "the frustratingly fleeting clues" to how sun-weather connections happen will lead, not too many years from now, to "practical and routine benefits to everyday forecasting."

Twenty-five to thirty years from now, he points out, we will inevitably have a nearly doubled world population. Needs for food will be acute, and climate and weather fluctuations will continue to pose enormous uncertainties in planning and growing the needed crops. "In the light of these considerations nothing could be more challenging than the effort to understand new mechanisms affecting weather and climate."

A dramatic new stimulus to sun-climate research came in 1976. Astronomer Jack Eddy's work—as I described in chapter 3—had confirmed that the sun's activity as determined by records of sunspots, auroras, and the tree-ring radiocarbon record, had been erratic in the past. For long intervals of time the sun, for whatever reason, apparently switches off some of these kinds of activity. The sun had been shown to be inconstant.

But there was something else as well. Eddy and others noticed that the two pronounced minima in the sun's activity in the past four centuries—the Maunder minimum and the Spörer minimum—happened at almost exactly the same time as the two coldest dips in climate in what we now call the Little Ice Age. These were times of severe winters and abnormal cold in Europe in the late 1400s and throughout most of the 1600s.

Climate change is a strange phenomenon. Climate does gradually vary, but the changes occur just beyond the time scale of human perception. The changes take place too slowly for us to be aware of them at the time. It is like trying to see the hour hand of a clock move. We cannot actually see the movement. We can only tell, after a long enough interval, that it *has* moved. Human society generally exists oblivious to changes in climate that are taking place over several human generations. Only later, when records of long enough duration have been compiled and analyzed, do we realize that a change has taken place. Yet the two cold dips comprising the Little Ice Age are real enough. Glaciers in Alaska, Scandinavia, and the Alps advanced close to their maximum positions since the last major ice age thousands of years ago. During the first of the two dips, expansion of the Arctic ice pack isolated the Norsemen's southwest Greenland colony and caused its inhabitants to perish. In Iceland grain that had grown for centuries could no longer survive. Historical records show that winters in both Paris and London were especially cold during the two periods. The Thames frequently froze for Queen Elizabeth's winter carnivals. Chemical analysis of ice cores from Greenland also confirms the two main dips.

The same kinds of climate records also show a pronounced "climate optimum," a prolonged period of unusually mild weather, from about A.D. 1100 to 1400 (except for a somewhat cooler period around 1200). This warm era coincides with the Medieval maximum in solar activity identified by Eddy.

Eddy referred to these correspondences as "close" and "striking." He found the correspondence no less striking when the record of earlier solar history was compared to the available records of earlier climate history, as best known, back several thousand years.

"The correspondence, feature for feature, is, I think, almost the fit of a key in a lock," said Eddy in a paper at a symposium on solar-terrestrial physics in Boulder in mid-1976. "Wherever a dip in solar activity occurs the climate swings coldward, and glaciers advance. When we have a prolonged maximum of solar activity glaciers retreat and the earth warms."

Eddy tentatively suggested that perhaps solar luminosity was slowly varying. These slight long-term changes in the solar constant wouldn't be *caused* by the prolonged sunspot absences; just the opposite. Some unknown changes in the solar dynamo would be responsible for both. The important indicator was not the eleven-year sunspot cycle (which might be totally irrelevant) but the long-term variations in the strength of many consecutive sunspot cycles, the so-called envelope of solar activity, showing its behavior over periods of fifty to one hundred years.

The identification of the Maunder minimum (and of the earlier ups and downs in solar activity) was considered certain and incontrovertible. The correlation with past climate fluctuations was intriguing but more speculative. The suggestion that solar luminosity was the explanation of the apparent associations was admittedly highly speculative.

I mentioned in chapter 3 that Eddy is both an imaginative and a cautious scientist. He has emphasized that the climate fluctuations used in these studies may reflect only regional (European, mainly) climate changes that may not be typical of the globe as a whole. He has pointed out that the climate variations were based on very coarse data. (Amazingly, as a result of the tree-ring radiocarbon record, the recent history of solar variation is more reliable than our climate records.) He acknowledged the possibility that the apparent "lock-and-key" fit of the solar and weather data might just be coincidence. Simple associations are always perilous. As one colleague warned, the Maunder minimum of 1645 to 1715 also exactly coincides with the reign of Louis XIV. "Could we say that a prolonged sunspot minimum produces a Sun King?"

Eddy had always been skeptical of many of the proposed

short-term links between sun and weather. But longer-term connections on the scale of decades, centuries, and millennia seemed more probable. The new evidence that climate might change in step with excursions in solar activity lasting fifty to several hundred years was intriguing. It stimulated unprecedented scientific and public interest in sun-climate connections in the late 1970s.

Yet, as Eddy wrote in 1978, "The connection seems exciting and important, but so did the false correlations formed in the 1870s." What must be understood, he said, is the real nature of the apparent relationship, what, if anything, really is happening to account for the presumed relationship. Better understanding of the sun and a much more finely detailed record of earth's climate variations were essential.

When I visited him in 1980, Eddy was less certain about his suggested sun-climate link than he had been originally. He pointed out that it always had been "the weakest part" of his work on the Maunder minimum. More recent work was causing some questions on how well the keys really did fit the lock. "As the climate record gets better defined," he said, "it's more complicated than we thought. I am getting concerned that the Maunder-minimum correlation with climate was oversimplified."

The new source of doubt was a study that Minze Stuiver was just about to publish in *Nature*. Stuiver had developed a way to get a much more finely detailed record of solar variation out of the carbon-14 content of tree rings. He had already, in work published in January 1980, used it to show the reality of the Maunder, Spörer, and Wolf sunspot minimums of the past eight hundred years (chapter 3). Now he had compared his fine-scale carbon-14 record of solar change with a variety of records of climate change. He had found fairly little correlation.

The closest agreement with the carbon-14 solar record was found in the index of the severity of winters in Russia. That coincided fairly well. But most of the other climate indexes did not. For instance, the coldest interval in Switzerland came not during either the Maunder minimum or Spörer mimimum but right between the two, from about 1550 to 1650. And Stuiver found that if the sun was responsible for climate fluctuations such as the Little Ice Age in Europe during the seventeenth century, then there should be many more Little Ice Ages than are seen in the climate record, perhaps as many as thirty or forty in the past 10,000 years. That is because the carbon-14 record shows the solar "Maunder mode" is quite prevalent. All in all, he found no statistically significant relationship between the combined regional climate and his high-precision carbon-14 record of solar activity.

"I don't want to claim that there is nothing to it," Stuiver told me, referring to Eddy's suggested sun-climate association. "Part of the problem may be in the climate record. Maybe there is an association in a few select regions. I don't want to say that the sun has nothing to do with climate change, only that it's not a simple problem." How true.

Future years should see the matter settled. There are real hopes for deciphering a record of solar variability extending back not thousands but millions of years from newly tapped kinds of earthly indicators, such as polar ice cores. Scientists hope to unravel that code and compare it with the long-term record of climate changes. Then the fit of solar variation and climate variations on the scale of fifty years to several centuries should, if real, be unambiguously revealed.

Climate varies over many scales. So also the sun. Eddy's work has examined the possible sun-climate connection on the scale of the reigns of kings or significant eras in human history. Workers such as Walter Orr Roberts have been concerned with possible sun-weather connections on the scale of a single human generation or even a single week. We'll look at the most recent findings in chapters 10 and 11.

First, however, in the next chapter, I tell the story of the recent verification of a sun-climate effect on the scale of the Ice Ages. This one is due not to any change in the sun itself but to cyclic changes in our planet's orientation *to* the sun. It explains in elegant fashion the major advance and retreats of the ice sheets that have alternately covered and exposed vast areas of our continents over the past million years—throughout all the long dawn of human prehistory.

9. DANCE OF THE ORBITS

One day in 1911, a young Yugoslavian poet and his thirty-two-year-old friend, an engineer and mathematician named Milutin Milankovitch, were in a Belgrade café celebrating the publication of the poet's book of patriotic verse. Coffee was all they could afford. But by good fortune, a well-dressed gentleman who turned out to be a bank director and a fervent patriot sat with them and began reading the poetry. He was so moved he asked for ten copies of the book and paid for them right there.

Now the two young men had something to celebrate with, and the red wine began to flow. By the third bottle, Milankovitch recalled later, their confidence was soaring and they had all the self-assurance of Alexander the Great looking for new worlds to conquer.

Emboldened by the spirits, the poet resolved to write an epic novel about the life and soul of their people. Milankovitch would not be outdone: "I feel attracted by infinity," he said. "I want to do more than you. I want to grasp the entire universe and spread light into its farthest corners."

The two friends parted happily. Milankovitch, inspired by the stimulus of that day in the coffeehouse, had found the challenge he had been searching for, a "cosmic problem" worthy of his greatest efforts. He would, he told his puzzled colleagues at the University of

Belgrade, develop a mathematical theory that would enable him to calculate the past and present temperatures and climates of the earth and its two nearest planets, Mars and Venus, at all latitudes and times.

His mathematical approach would not be limited by the irregular distribution of weather stations. He would be able to investigate temperatures anywhere—over the oceans, high up in the atmosphere, or even on the other planets. And he would be able to peer into the past and determine what previous climates must have been like when the shape of the earth's orbit, the tilt of the earth's axis, and the seasonal distances between the sun and earth were different than they are today. He would put the astronomical theory of the ice ages on a firm mathematical basis, something no one else had been able to do. The quest would take thirty years.

Although no one had yet calculated the magnitude of all the effects accurately, a remarkable amount of work on earth's orbital variations had already been done. There had even been considerable discussion on their possible effects on climate change and the comings and goings of the ice ages.

It was well known, for example, that the earth's orientation in space does not always stay the same. Way back in ancient Greece in 120 B.C., Hipparchus, comparing his observations to those made by another astronomer 150 years earlier, noticed that the direction in the night sky the earth's North Pole points to had slowly changed. Plotting all this out on a star map, astronomers were able to show that the earth was slowly wobbling, like a spinning top. It turns out that the earth completes one such wobble every 25,800 years. Our planet's axis thus describes a circle about 47° diameter in the night sky over that period of time. Earth's North Pole now points roughly to Polaris, the pole star, at the end of the handle of the Little Dipper. But in 2000 B.C., the North Pole pointed to a spot midway between the Little Dipper and the Big Dipper. And in A.D. 14,980 the North Pole will point to Vega, the bright star now nearly directly overhead in the summertime sky.

This phenomenon, called axial precession, was explained by a French mathematician, Jean Le Rond d'Alembert, in 1754. He showed that it was due to the gravitational pull exerted by the moon and sun on earth's equatorial bulge.

Axial precession causes the four cardinal points to move

slowly around earth's orbital path in a clockwise direction (as seen from an observer looking down from our north). At the same time the earth's elliptical orbit is itself, even more slowly, rotating in space in the opposite direction. That motion cancels out some of the effect of the axial precession. The combined result is a cycle lasting 22,000 years known as the precession of the equinoxes.

The result of all this is that at a given season the distance between the earth and the sun slowly varies. Imagine the earth's elliptical orbit as the cross section of an egg.* The sun is inside it, slightly closer to the large end of the egg than to the smaller end.

Let's take a few examples. Winter in the northern hemisphere begins when the earth is at the position in its orbit where the northern half of the planet is facing its maximum possible extent away from the sun. That is what we mean by the beginning of winter. It is usually December 21 on our calendar.

In today's era, it so happens that winter in the northern hemisphere begins when the earth is near the large end of the egg, in other words when it is near its closest point to the sun. Northern hemisphere summer, on the other hand, begins very near the time when the earth is at the small end of the egg, near its farthest point from the sun.

Now, the amount by which the earth's orbit varies from a circle is not very much, only about 1.67 percent. But it's enough to mean that near the beginning of northern hemisphere winter the earth is 5 million kilometers, or about 3.3 percent, closer to the sun than it is near the beginning of northern hemisphere summer.

The precession of the equinoxes, however, causes this situation gradually to shift over time. Eleven thousand years ago (as well as 11,000 years from now) winter in the northern hemisphere began when the earth was near its *farthest* point from the sun. Summer began when the earth was near its *closest* point to the sun.

Obviously this means that the seasonal distribution of sunlight slowly changes over this 22,000-year cycle too. In the present era, the earth gets a little more heat from the sun during winter in the northern hemisphere and a little less heat during winter in the southern hemisphere than the long-term average. Summers are just the opposite.

I know the earth's orbit is a true ellipse and not really egg-shaped, but the analogy is close enough for our present purposes.

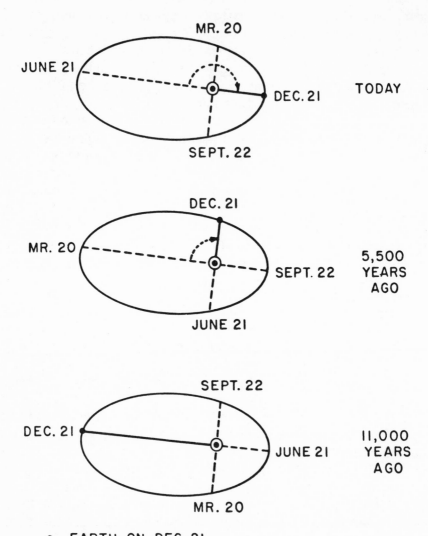

Precession of the equinoxes. As a result of precession of the earth's axis and other astronomical influences, the point at which a particular season begins shifts slowly around earth's elliptical orbit, completing one full cycle in about 22,000 years. At present, northern hemisphere winter begins when the earth is near its closest point to the sun. Eleven thousand years ago it began when the earth was near its farthest point from the sun. This is one of the three important orbital influences on climate change. (John and Katherine Palmer Imbrie)

During summer, the earth receives a little less heat from the sun in the northern hemisphere and a little more in the southern hemisphere than the long-term average.

Twenty-two thousand years ago, that was all reversed.

In 1842, in a book called *Revolutions of the Sea*, a French mathematician named Joseph Alphonse Adhémar theorized that glacial climates come and go as a result of this 22,000-year cycle. He didn't get too far with the idea, however. This was partly because he muddled the matter by proposing some exotic but fallacious scenarios involving the gravitational attraction of the huge Antarctic ice sheet for the water of the northern hemisphere oceans. Also, scientists pointed out correctly that these precession effects all evened out. Any increase in heat because the earth is closer to the sun at one season is exactly balanced by its being farther from the sun at the opposite season. Throughout the year, the total amount of sunlight reaching earth is constant no matter what part of the 22,000-year precession cycle we're in. And that is true for each hemisphere as well.

Yet there was something else to consider. The axial precession and slow rotation of earth's elliptical orbit weren't the only influences that changed earth's long-term orientation to the sun. The great French astronomer Urbain Jean Joseph Leverrier showed in 1843 that the *shape* of earth's orbit changes. Since Kepler's time the orbit had been known to be elliptical, but Leverrier demonstrated that the degree of elongation of the ellipse, usually known as eccentricity of the orbit, changes slowly and continuously over a roughly 100,000-year period.

This, he showed, was a consequence of the law of gravitation formulated by Newton. Each of the planets as it revolves round the sun exerts a gravitational influence on earth's orbit. It took ten years, but Leverrier managed to calculate the effects of all these ever-varying gravitational influences of the known planets.* They showed that the shape of the ellipse that defines earth's orbit is constantly changing. He found that the amount of eccentricity—the degree to which the orbit is out of round—varies from as much as 6 percent to as

---

*This work led Leverrier to another triumphal application of Newton's theory of gravitation. In 1846 he successfully predicted the existence and location of a new outer planet based on its effects on Uranus's orbit. Astronomers trained their telescopes on that spot and found it. The new planet was almost called Leverrier, but Leverrier named it instead after the Greek god of the ocean, Neptune.

little as nearly zero (almost round) over 100,000 years. (It is now, as I mentioned, 1.7 percent.)

It was an astonishing thought that even the shape of earth's orbit changes regularly. It meant that the distance from the sun to the earth, which now varies only slightly throughout one orbit, had at times in the past gone through both more pronounced and even less pronounced annual variations. It seemed that might influence climatic change.

Enter James Croll, a Scottish mechanic turned self-taught natural scientist.

Croll, born in 1821, had a philosophical mind that ventured far beyond the mundane practical pursuits he took on to subsist. By the late 1850s he had worked as a millwright, a carpenter, a tea-shop owner, a hotelkeeper, and an insurance salesman. In his teens he had developed a passion to understand the physical laws of nature, and in his thirties he published a well-received philosophical treatise.

A job he took as a janitor at Andersonian College and Museum in Glasgow in 1859 gave him leisure and access to a good scientific library. He turned both to good use. In the early 1860s geologists were speculating on the causes of the glacial age, whose reality had only recently been firmly accepted.

Croll turned to the question. He felt certain there was an astronomical cause. Adhémar had gone wrong, he reasoned, by considering only the earth's wobble and the precession of the equinoxes. Adhémar had not taken into account Leverrier's discovery of the changing elongation of earth's orbit. The change in orbital shape was the missing key, he felt.

Croll used Leverrier's formulas to calculate the change of orbital eccentricity over particular past intervals. He discovered that intervals of high eccentricity had peaked roughly 100,000, 200,000, and 300,000 years ago and that the most recent peak had been followed by a long decline to an unusually low eccentricity (nearly circular orbit) about 10,000 years ago.

This seemed to him likely to have something to do with ice ages.

The variations in orbital eccentricity did not affect the total amount of sunlight reaching earth throughout the year, of course. (There was no change in the average sun-earth distance.) But Croll showed that the amount of sunlight reaching earth each season was strongly affected.

Observing the sun and sunspots in the early seventeenth century.
Here Christoph Scheiner and a fellow Jesuit scientist, in Italy about
A.D. 1625, trace sunspots projected through a telescope onto a screen.
(From Scheiner's Rosa Ursina, *by permission of Houghton Library, Harvard University)*

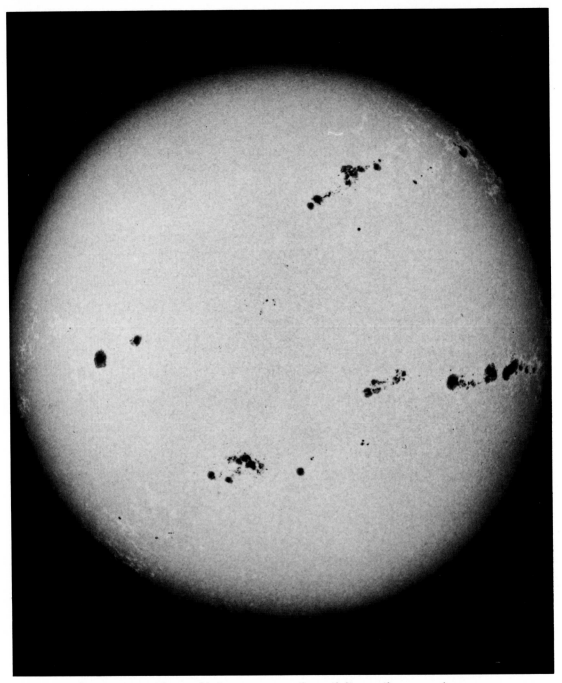

Sunspots, regions of intense magnetic activity on the sun, show up darker than surrounding areas when seen in ordinary light because their temperatures are somewhat cooler. *(Mount Wilson and Las Campanas Observatories, Carnegie Institution of Washington)*

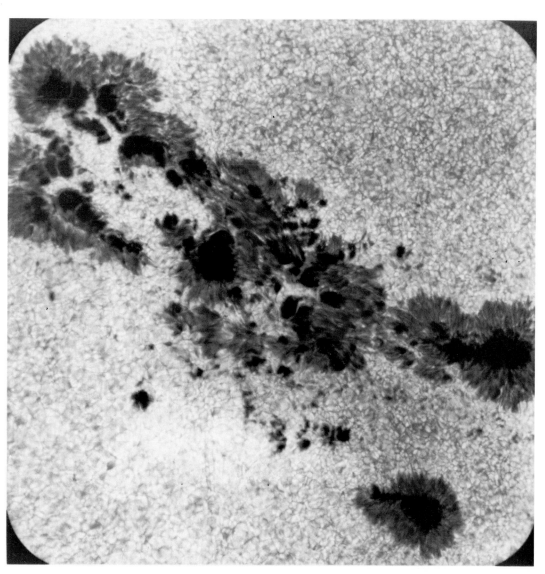

Complex sunspot group, showing dark central umbras surrounded by delicate, filamentary penumbras. *(Sacramento Peak Observatory, © 1979)*

The shimmering aurora borealis, the beautiful manifestation on earth of storms on the sun. *(Courtesy S.-I. Akasofu)*

1973

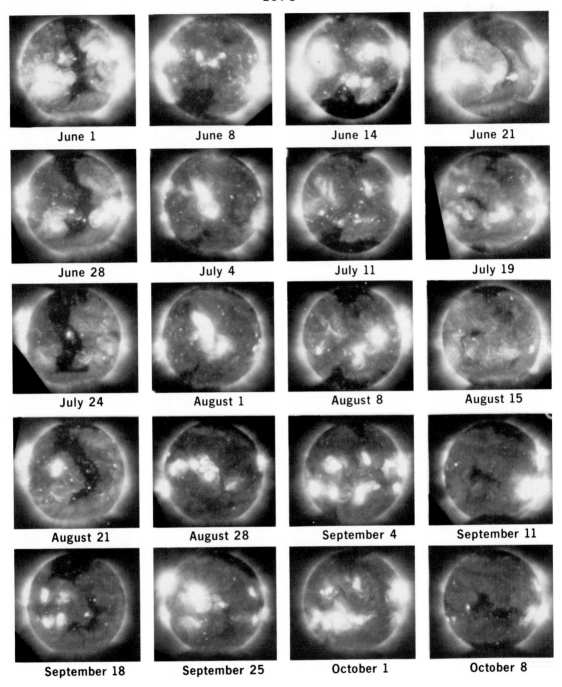

June 1 June 8 June 14 June 21

June 28 July 4 July 11 July 19

July 24 August 1 August 8 August 15

August 21 August 28 September 4 September 11

September 18 September 25 October 1 October 8

X-ray photos of the sun taken at one-week intervals. They are arranged so each vertical column shows the sun at 27-day intervals (the rotation period of the sun). The left-hand column shows development and life history of large coronal hole, the dark area extending down across the sun's equator. Coronal holes were shown to be the source of the high-speed solar wind streams producing mysterious 27-day recurrent geomagnetic disturbances on earth. *(American Science and Engineering, Inc.)*

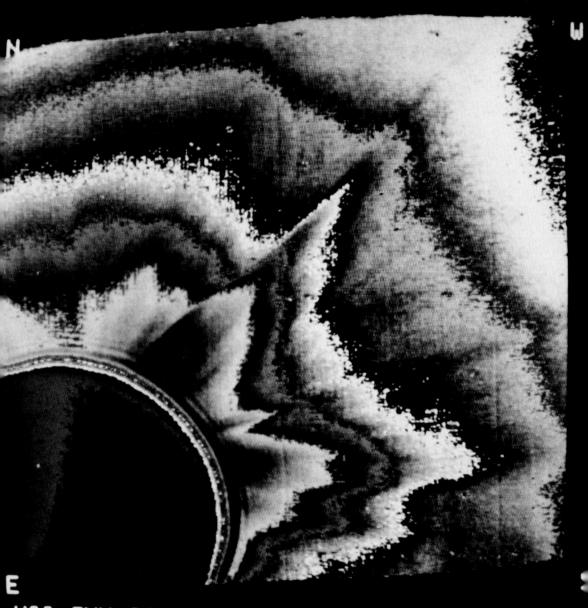

N

W

E

HAO SMM CORONAGRAPH/POLARIMETER
DOY 103   UT= 1416 POL=0

Striking view of the solar corona in mid-April 1980 from data supplied
by the Solar Maximum Mission satellite. Computer shading represents
different densities in the corona. The largest of the several coronal
spikes persists beyond a million miles from the sun's surface. *(NASA/High
Altitude Observatory)*

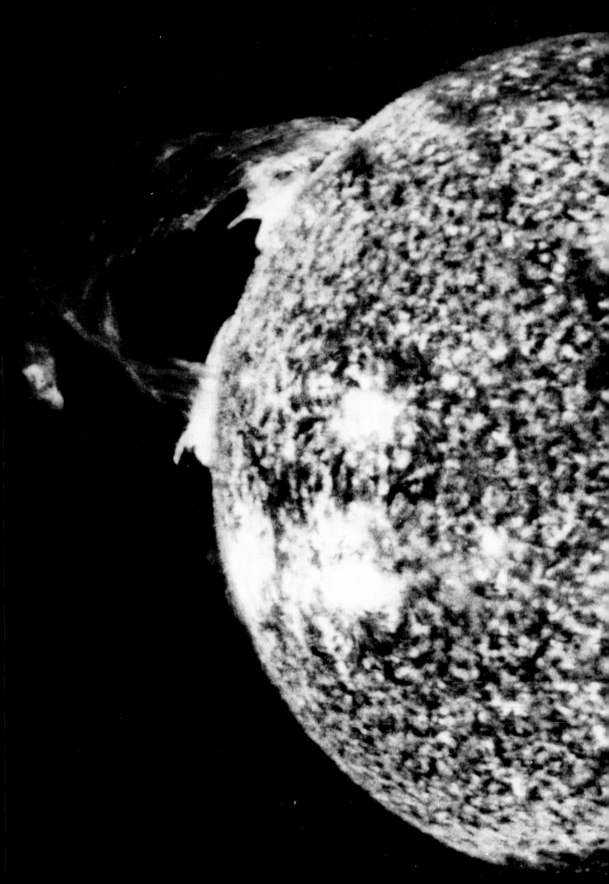

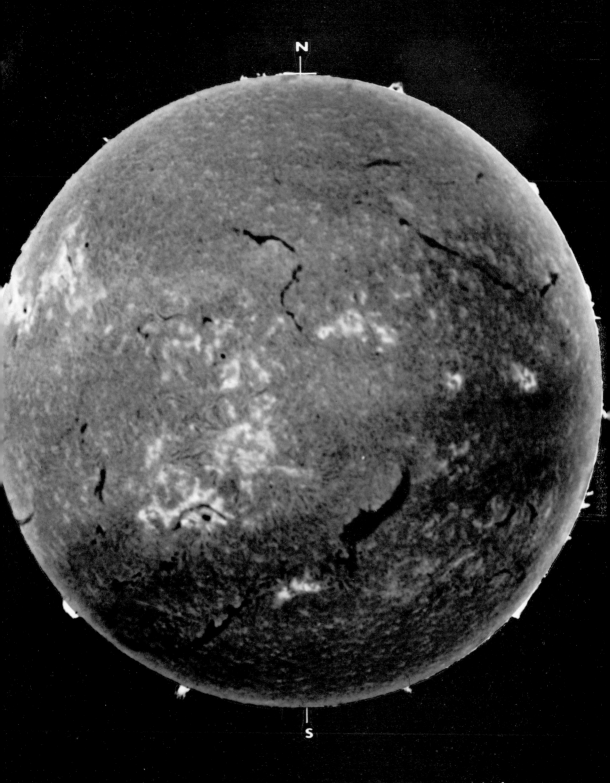

Hydrogen light photo of sun on September 28, 1980, near peak of
the current solar cycle. *(Space Environment Services Center, NOAA)*

Painting of warped heliospheric current sheet. As entire structure rotates with sun over 27-day period, one of its surfaces transits earth roughly every week. *(Courtesy John M. Wilcox; artist: Werner Heil)*

Trajectories of the two originally planned solar-polar spacecraft. The U.S. has canceled its solar-polar spacecraft, leaving the European craft to go it alone. *(NASA/Jet Propulsion Laboratory)*

Proposed solar probe mission would take a spacecraft within several million miles of the sun's surface. *(NASA/Jet Propulsion Laboratory)*

Charles Greeley Abbot, here with one of his solar devices outside the Smithsonian Institution, pioneered the attempted long-term measurement of variation of the solar constant. *(Smithsonian Institution)*

Andrew Ellicott Douglass, founder of dendrochronology, thought that tree rings would show a clear effect of the 11-year sunspot cycle on weather and climate. *(University of Arizona)*

Walter Orr Roberts came west to run the High Altitude Observatory and eventually became a strong if philosophical proponent of sun-weather effects. *(National Center for Atmospheric Research)*

John A. Eddy confirmed and named the Maunder sunspot minimum and stimulated unprecedented new interest and research into all aspects of solar variability. *(National Center for Atmospheric Research)*

Henry A. Hill identified global oscillations on the sun. *(University of Arizona photo by George Kew)*

Richard C. Willson inaugurated the new era of solar constant measurements using instruments on balloons, rockets, and the Solar Maximum Mission satellite. *(Jet Propulsion Laboratory)*

Philip H. Scherrer and John M. Wilcox, here at the Stanford Solar Observatory, studied solar oscillations and elucidated the solar sector structure and its possible effects on weather patterns. *(Stanford University)*

How did it all work? Croll believed that ice ages were produced when winters were especially cold. That would happen during eras when winter occurred (in the northern hemisphere, say) while earth was at a greater than usual distance from the sun. The precession of the equinoxes governed the timing of those situations, but the variations in orbital eccentricity governed their intensity. If, for example, the orbit was circular, the precession of the equinoxes would have no effect on the amount of sunlight received in a given season. It would always be the same. But at times of higher elongation the precession effect would go to work. Then the seasonal distribution of sunlight would change according to whether the earth was in a part of its orbit farther or nearer the sun than usual.

The orbital events seemed to fit the glacial facts. When the earth's orbit was nearly circular, as it had been for the past 10,000 years, winters would be average. No ice age could take place. But in earlier times, ice ages could occur whenever winter occurs when the earth is farthest from the sun and when the eccentricity variations made that farthest point even more distant than usual.

Eventually Croll modified the theory further. Its final form, published in his 1875 book *Climate and Time,* took into account a third variation in sun-earth orientation. In addition to precession and orbit-shape changes, it turned out that the tilt of earth's axis varies slightly over time.

The tilt of the axis is all-important. It is the fundamental cause of seasonal change. Earth's axis is now tilted 23½° from the vertical, but Leverrier's calculations had shown that this amount could vary over time between 22° and 24½°. The effect of this was not totally clear to Croll. He reasoned that ice ages would be more likely when the tilt was closer to the vertical, because then the polar regions would receive less sunlight. But Leverrier had not calculated the timing of these variations in tilt, so Cross was unable to do much with them. (We now know the angle of tilt varies over a cycle of 41,000 years.)

Croll's theory stimulated enormous debate in the community of geologists and meteorologists considering the causes of the ice ages. He had both strong supporters and detractors. The basic problem was that the geological record of the ice ages was not good enough to settle the matter. His work became increasingly vulnerable to the accumulating evidence that the ice sheets had swooped down across the northern continents far more recently than the 80,000 or 100,000 years ago his orbital-eccentricity theory indicated. New evidence from North America showed that the ice sheets had left as recently as 6000 to 10,000 years ago. Furthermore the effects just didn't seem large enough to account for such a striking phenomenon as a worldwide glacial advance.

By the end of the nineteenth century, the astronomical hypothesis for the ice ages had been put aside and nearly forgotten.

All this Milankovitch learned when, in his methodical way, he studied the past history of this quest before embarking on it on his own. Croll and the others, he realized, hadn't had the mathematical training to do what he intended.

First he found that by good fortune a German mathematician named Ludwig Pilgrim had in 1904 done all the necessary orbital calculations. Whereas Croll had had to make do with calculations of precession and eccentricity back 100,000 years, Milankovitch now had at his disposal calculations of those two orbital parameters plus the tilt of earth's axis back a million years.

He then applied himself to calculating how all those varying influences affected the amount of solar radiation striking earth (and also Mars and Venus) during each season and at each latitude. It was a formidable mathematical task. As Milankovitch said later, had he been younger he wouldn't have yet had the knowledge and skills to accomplish it and had he been older he would have foreseen too many obstacles to even start it. At thirty-two, he had the right combination of unfettered enthusiasm and well-tested ability.

Milankovitch spent some part of each day on the work, even family holidays, most of it in his book-lined study in his Belgrade home. Within a few years, despite brief service in the Balkan War in 1912, he had significant results. His 1914 paper, "On the Problem of the Astronomical Theory of the Ice Ages," was written in Serbian and largely ignored. Yet it demonstrated mathematically that the variations in axial precession and orbital eccentricity were indeed large enough to account for the advances and retreats of the ice sheets. And it showed that the effect of the changes in tilt of earth's axis were even more important than Croll had suspected.

His work was again interrupted by World War I. He was even held for a time in 1914 as a prisoner of war after being captured by the Austro-Hungarian Army while on a visit to his hometown. A mathematician friend successfully petitioned for his release in the interest of science, and Milankovitch was sent to Budapest, where he had to report to the police once a week. He settled in at the Library of the Hungarian Academy of Sciences and for four more years continued his work.

The book he eventually published in 1920 describing all this work caught the attention of a respected German climatologist named Wladimir Köppen, and the two scholars began a productive correspondence. Milankovitch was now ready to turn his full attention to the mathematical description of past climates on the basis of the orbital variations. He wasn't certain that Croll had been right in assuming that the condition most conducive to ice ages was unusually cold winters. He asked Köppen about it, and the climatologist replied that he considered the most important factor to be not cold winters but cool summers. Winters would always produce abundant snow in the high latitudes. But cool summers, Köppen reasoned, would inhibit the melting of snow and glaciers in the high latitudes and high elevations so the next winter's onslaught of snow would cause them to grow. This realization would prove to be of fundamental importance.

Milankovitch at once turned to calculating the summertime solar radiation curves for latitudes 55°, 65°, and 75° on earth for the past 650,000 years. This too wasn't easy. He worked on these calculations for one hundred days, then charted the results on a graph and mailed it off to Köppen.

Köppen, who was in the midst of work on a book about past climates with his son-in-law Alfred Wegener,* wrote back immediately. The patterns of cycles in Milankovitch's radiation curves fit reasonably well with the major fluctuations of past climates as determined by the history of alpine glaciers. For instance, the Wurm, Riss, Mindel, and Gunz ice ages of the past 600,000 years all seemed to coincide with eras when summer solar radiation was at a low point.

Köppen included Milankovitch's charts in his and Wegener's climatology book, published in 1924, and that caught geologists' attention. In a scientific meeting in Innsbruck, Wegener spoke so highly of the Yugoslavian's work that Milankovitch found himself "quite embarrassed."

Nevertheless, he was very pleased.

Wegener is best remembered now as the originator of the theory of continental drift, now grandly confirmed in a more sophisticated version as the concepts of sea-floor spreading and global plate tectonics. But originally he was a meteorologist.

In succeeding years, Milankovitch calculated past solar radiation curves for five more bands of latitudes, so now everything from 5°N to 75°N was included. All this work was published in a single volume in 1930. Geologists could now see how all these orbital parameters affected the amount of solar radiation each latitude of earth had received in the past. Each of the three astronomical factors had their most pronounced effects at different latitudes. The 22,000-year precession effect, for instance, exerted its largest influence at the equator and its smallest at the poles. The 41,000-year cycle of variation in the tilt of earth's axis (with greater tilt resulting in more summer radiation) was just the opposite. It exerted its largest influence at the poles and its smallest at the equator.

Milankovitch finished up all this work with the publication in 1941 of a book applying all these radiation variations to what was known about their complex effects on ice sheets. He managed to develop a way, for instance, to estimate the effect more extensive ice has on temperatures due to its greater reflectivity of sunlight. These kinds of "positive feedbacks" amplify the effects of reduced solar radiation caused by the astronomical variations. A global cooling trend, as a result, becomes far more pronounced than it would otherwise be. The same with a warming trend.

It had taken three decades of painstaking work from the time Milankovitch sat with his poet friend in that Belgrade coffeehouse vowing to take on a cosmic problem. To his imminent satisfaction, that problem had been solved.

Yet it wasn't that simple. The geological record of the ice ages was still woefully inadequate. The evidence to test the Milankovitch theory was fragmentary and often contradictory. The most solid evidence for the timing of recurrence of ice ages showed expansions and contractions at roughly 100,000-year intervals. Yet Milankovitch's calculations of axial-tilt and precession cycles showed that there should be even more significant climate variations at 41,000-year and 22,000-year intervals. Climate cycles of that duration just didn't seem to show up in the imperfect records available.

What was now called the Milankovitch astronomical theory of the ice ages steadily lost supporters, and by 1965 it seemed once again to be back where Croll's detractors had left it in the late

1800s—a curious theory having some interesting historical value but scientifically virtually a dead issue.

But the scientific revolution that ten years later would cause it once again to rise, phoenixlike, from the rubble of discarded ideas was already in progress. The tools were coming forth that could provide the fine details about past climate fluctuations, a kind of microscopic record of the past, equal to the task of testing the Milankovitch theory in all its rhythmic elegance.

Wallace S. Broecker of Columbia University and Robley K. Matthews of Brown University investigated ancient coral reef terraces on oceanic islands for evidence of the pronounced rises in sea level that should accompany any anomalously warm era between glacial periods. Coral grows only near sea level. When the seas subside as more and more water is locked up in glaciers on land, a horizontal terrace of coral is left high and dry. A whole succession of such terraces can be left by different episodes of glacial advance. In the mid-1950s a technique had been developed to date the coral by use of thorium.

When Broecker, Matthews, and their graduate students applied this technique to the three coral terraces on the Caribbean island of Barbados they found them to be 82,000, 105,000 and 125,000 years old. The Milankovitch solar radiation curves for lower latitudes contained distinct maximums near those dates. The 22,000-year precession cycle seemed to be responsible. Other investigations on New Guinea and the Hawaiian Islands found similar results. This work, published in the late 1960s, began a revival of interest in the astronomical theory. In 1970, in fact, Broecker and a colleague reviewed the evidence for it and found that the case was now getting quite strong.

A marine paleontologist named Cesare Emiliani was responsible for much of this bolstered status. Emiliani had set in motion an exotic new way of measuring past climatic conditions. Stimulated by the ideas of the great Nobel laureate geochemist Harold Urey at the University of Chicago, where he had done his postgraduate studies, Emiliani looked to the oceans for the answer. The isotopic composition of oxygen that tiny marine organisms extract from the sea and lock up in their skeletons as calcium carbonate depends, Urey had theorized, upon the temperature of the seawater at the time. A temperature record of past ages should exist in the sediments of the oceans, graveyard of the sea's microscopic but phenomenally abundant life forms.

Emiliani had at his disposal an absolutely precious resource: hundreds of sea-floor sedimentary cores extracted almost daily since the early 1940s by the research ships of Columbia University's Lamont Geological Observatory on the instructions of its visionary director, Maurice Ewing. New techniques of weighing molecules had also become available.

Emiliani examined representative cores and found that the proportion of heavy oxygen (oxygen-18) to normal oxygen in the skeletal remains did indeed vary cyclically in patterns that seemed to indicate striking changes in the global environment. Times when the proportion of heavy oxygen was high were interpreted as cool periods; low, warm periods.

The result was a much more finely detailed record of past climate change than had ever existed before. There were more cycles of cold and warm conditions, and at shorter intervals, than had been previously suspected. The old idea of just four major ice ages was crumbling.

It wasn't until the 1970s that scientists came to realize that what Emiliani's oxygen isotopes were really measuring was not ocean temperature but a climatic indicator even more profound: the volume of glacial ice on land. A young Cambridge University geophysicist from England, Nicholas J. Shackleton, and a geologist-oceanographer from Brown University, John Imbrie, were primarily responsible for that insight.

It seems astonishing that the chemicals in the skeletons of tiny marine creatures buried beneath the bottom of the ocean can tell us the extent of the continental ice sheets at the time they lived. But they do.

Normal oxygen, with an "atomic weight" of 16, is lighter than the oxygen-18 isotope. Water molecules consisting of normal oxygen are therefore preferentially evaporated from the surface of the oceans. When the extent of the ice sheets is constant, the normal oxygen, which precipitates out of the air as rain or snow, soon finds its way back to the oceans. The sea composition doesn't change. But when the ice sheets on land are expanding, some of the oxygen-16 gets locked up in the form of ice and does not return to the sea. The proportion of the heavier oxygen-18 in the oceans then rises. The amount of oxygen-18 in the oceans is at a maximum at times of maximum ice volume on land and at a minimum at times of minimum glacial volume.

The microscopic marine skeletons Emiliani studied were indeed chemical messengers from the past. But their message was not temperature itself but ice extent. They were giving scientists a vivid re-created view of the advances and retreats of the ice ages.

The problem left now was to accurately time all these events.

In 1963, three American geophysicists had shown conclusively that the earth's magnetic field reverses its direction of polarity every million years or so and that the record of these reversals is frozen into lava that was cooling at that time. These flows could be dated by a technique known as potassium-argon dating. The time of the last reversal was thus known to have been 700,000 years ago. The one before that was 1.8 million years ago. A whole magnetic-reversal calendar was established.

Soon ocean-floor cores were found that contained both information on climate variations and signs of these magnetic reversals. Scientists thus had a way of identifying the exact times of the climatic variations, a way to calibrate their climate record.

By the early 1970s, it was apparent that deep-sea cores were ready to be applied to an all-out assault on the mystery of the ice ages. At last the astronomical theory of glacial advances and retreats could be given the kind of sophisticated test it deserved.

Shackleton and Imbrie soon joined up with James D. Hays of Columbia's Lamont-Doherty Geological Observatory to tackle the

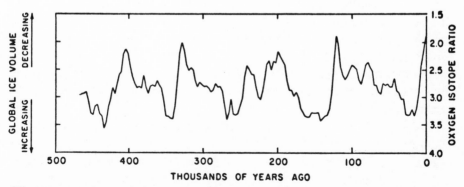

Climate variations of the past half-million years deduced from two Indian Ocean deep-sea sediment cores. These observations led to confirmation in 1976 of the astronomical theory of the ice ages. (John and Katherine Palmer Imbrie; data from Hays, Imbrie, and Shackleton)

question. Hays was a specialist on microfossils of tiny Antarctic ocean-water animals called radiolaria. He had a strong interest in applying the record of their variations in deep-sea sediment cores to climate history. The trio's work was actually one part of a larger national effort, called CLIMAP, to map the surface of the earth during the last ice age and to chart climate oscillations for the past million or so years.

The problem was to find a particular type of sedimentary core. It should have a high sedimentation rate—the rain of microscopic animal life and other debris from the surface waters should have been intense enough to accumulate at a rate of three millimeters per century. (A millimeter is about the thickness of the wire in a paperclip.) A lesser rate of accumulation would result in a fossil record too compressed to avoid being disturbed by the burrowing activities of animals on the sea floor.

It should contain shells of both radiolaria and the type of animal called foraminifera. The oxygen-18 composition in the latter would provide the record of fluctuations of the continental ice sheet, primarily a northern hemisphere phenomenon. Statistical studies of the populations of the former would provide direct evidence of local sea-surface temperature at the time the radiolaria had lived. Such studies can serve as a sensitive thermometer of the past because different species of radiolaria flourish at different temperatures. One particular species of radiolaria would provide still a third check. It was remarkably abundant during glacial times, apparently as a result of a particular temperature-salinity structure of the water that existed then.

The core should also be from high latitudes of the southern hemisphere far from any land to avoid any local effects.

Hays eventually found two cores that together fit the bill. One extended back only 300,000 years, but it could be patched together with the other to extend the record back 450,000 years. Both were from the southern Indian Ocean bottom about midway between the tips of South Africa and Australia.

The scientists carried out the detailed detective work necessary to decipher the climate record in the core. The radiolarian abundances and oxygen-18 quantities were carefully examined.

A clear record of climate variation was apparent.

The most prominent cycle was of roughly 100,000 years. That agreed with the findings in many previous core studies. But many shorter-term variations were visible too.

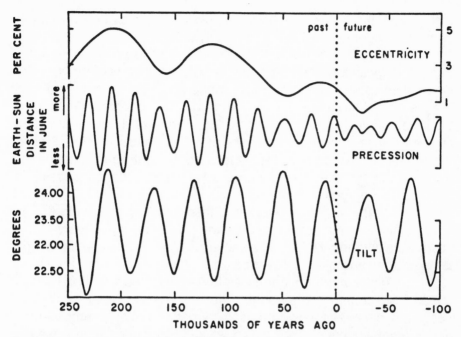

Changes in eccentricity, tilt, and precession. These changes in the geometry of the earth's orbit in relation to the sun have proved fundamental to climate change. (John and Katherine Palmer Imbrie; data from A. Berger)

After all sorts of sophisticated analysis the evidence was clear. There were pronounced climate cycles of 100,000 years, 43,000 years, and a twin cycle of 24,000 and 19,500 years. The latter was puzzling. Now came time to compare the climate results with the solar radiation curves of the astronomical theory. Anandu Vernekar of the University of Maryland had recently recalculated them all, and Imbrie got copies from him. They showed that the cycle of varying tilt of earth's axis is 41,000 years. That was just as Milankovitch and his predecessors had calculated. No surprise there. But the newly calculated curve of the precession effect showed two cycles of 23,000 and 19,000 years. Belgian astronomer André Berger confirmed this fact for Imbrie. To the scientists' enormous satisfaction these two previously unsuspected twin components of the precession cycle had shown up in the climate record from their deep-sea cores.

The overall agreement between the newfound climate record and the calculated astronomical curves was remarkable. Each of the cycles from the cores matched the predicted cycles to within 5 percent. The evidence was convincing.

Hays, Imbrie, and Shackleton published the full report of their findings as a twelve-page lead article in the December 10, 1976, issue of *Science.* Titled "Variations in the Earth's Orbit: Pacemaker of the Ice Ages," the article went through all the technical intricacies of their methodology and then concluded: "We regard the results of the time-domain test as strong evidence of orbital control of major Pleistocene climatic changes.... It is concluded that changes in the earth's orbital geometry are the fundamental cause of the succession of Quaternary ice ages."

All the variations were in phase and in the expected direction. The times of minimal tilt of the earth's axis that came every 41,000 years were invariably followed by an equivalent dip to a period of cold temperatures and glacial advance.

Times when the earth was farther from the sun than usual at the start of summer in the northern hemisphere as a result of the precession cycle were followed closely by periods of cold temperatures and glacial advance. That was just as Köppen had proposed to Milankovitch back in 1920. Before 300,000 years ago, this effect became less pronounced. That too might be expected on orbital considerations. From about 450,000 to 350,000 years ago, earth's orbit stayed very nearly circular, according to Berger's calculations, so precession could have little effect on the seasonal distribution of sunlight on earth.

In all these cases glacial advance lagged behind cooling temperatures, just as would be expected. The big surprise was the strong, apparently direct effect of the roughly 100,000-year cycle of orbital eccentricity.* Previous studies had predicted that the varying shape of the orbit should affect the climate record mainly as a modulation of the 22,000-year precession cycle. But Hays, Imbrie, and Shackleton showed without a doubt that there is some direct effect. Times when earth's orbit was most nearly circular were periods of cold and glacial advance.

---

*The eccentricity doesn't change by the same amount from cycle to cycle. The time for a cycle varies too. Berger's calculations show that over the past million years the eccentricity has varied between 0.057 percent and 5.4 percent with the cycles averaging 95,200 years. Over the next million years, it will vary from 0.27 percent to 5.7 percent with the cycles averaging 91,444 years. We are now at the beginning of a period of 101,000 years during which the eccentricity remains small.

Why should that be? Well, suppose that in the present configuration of the continents—most of the land is in the northern hemisphere and is clustered around an Arctic sea—glacial conditions are the normal situation. In that case, the question becomes not "Why does an ice age start?" but "What causes the ice to temporarily recede?" When the orbit is circular, the earth's precession really has little chance to change a given season's receipt of solar radiation. But a more elliptical orbit results in periods when summer in the northern hemisphere can coincide with minimum distance to the sun—as it did about 11,000 years ago. That extra injection of summertime heat—especially if combined with a greater tilt of the axis, causing the sun to be slightly higher in the sky—may be just enough to put the ice sheets into retreat—for a while. Then it's back to the ice and cold again.

But wait a minute. It's only been for the past two or three million years—the Pleistocene period—that the earth has seen ice sheets repeatedly cover the northern parts of the continents. For some 250 million years before that there is no sign of ice ages. Yet presumably all the orbital cycles we've been invoking to explain the ice ages were operating throughout that long era of earth history too. Why weren't there ice ages then as well?

The answer may well be the changing positions of the continents around the globe. It turns out that the present configuration of the northern continents is indeed very conducive to periodic ice ages on them. A large proportion of the land surface of earth is in the northern hemisphere, and the three northern continents, North America, Europe, and Asia (plus Greenland, which is covered by an ice sheet now), all crowd up around a polar ocean. The polar ocean supplies plenty of moisture for snowfall in the northern regions, and the continents supply plenty of land area on which it can accumulate. The stage is set. Given the proper sun-earth astronomical setting to produce a period of cool summers and warm winters, an ice age can begin. Warmer-than-usual winters in the Arctic north allow more than the usual amount of snow to fall (air that is too cold can carry little moisture). More important, cool summers keep some of the previous winter's accumulation from melting. This continues for many seasons and soon the amount of land covered throughout the year by snow is expanding. This triggers even more ice expansion because the white snow and ice reflect a large fraction of the sunlight and further cool the summers. The colder summers also cause a change in vegetation patterns, forests giving way to tundra. Snow-covered tundra is much more reflective than snow-covered forests. These "positive feedbacks" accelerate the trend. An ice age is under way.

We know now that the continents are in slow continual motion—they change their position on the globe by about the width of your finger each year. For many millions of years the general direction of their drift has been northward. Until about three million years ago, then, the northern continents weren't up quite so tight around the polar ocean, and the conditions presumably weren't right for the astronomical variations in earth's seasonal distance and angle to the sun to initiate cycles of ice ages. With the continents now in their high-latitude position, we appear to be locked into an ice-age condition, with orbital variations stimulating a series of glacial advances and retreats.

Where does the astronomical theory of the ice ages say we are headed now? The tilt of our planet's axis and the eccentricity of our orbit around the sun are both decreasing. Those two trends favor a return to ice age conditions. Precession is beginning to take us toward a time of shorter sun-earth distance in summer, a condition that discourages ice ages. To find the net result of all three effects, Imbrie has developed a mathematical formula based on Berger's calculations of solar radiation curves. The formula indicates that on astronomical considerations alone, the global cooling trend that began 7,000 years ago will continue and eventually bring us to the next ice age in 7,000 or 8,000 years, reaching the maximum glacial advance 23,000 years from now. Our orbital geometry is taking us toward steadily cooler climate.

Yet many other influences are at work too. The warming effect of all the carbon dioxide our civilization is putting into the atmosphere may well temporarily cancel out and even overcome the cooling in the coming several centuries. And shorter influences of uncertain cause, such as whatever stimulated the Little Ice Age and similar intervals of cold or warmth on the scale of a century or so, are at work as well. Nevertheless the astronomical theory of changing sun-earth orientation gives a clear and predictable view of inexorable trends that affect our planet's climate on the scale of tens of thousands and hundreds of thousands of years.

Most scientists have in the past five years accepted the validity of the astronomical theory of the ice ages. The concept has gone through its own cycles of acceptance and disrepute since the pioneering work of Croll and Milankovitch. But Hays, Imbrie, and Shackleton, drawing on the fossil and chemical record of the ocean bottom, finally produced the decisive test. The title of the excellent book* Imbrie and his daughter Katherine wrote chronicling this long search sums it up well: *The Ice Ages: The Mystery Solved.* Climate fluctuations on the scale of tens of thousands to hundreds of thousands of years definitely seem to have their origin in our planet's geometrical relationship to the sun. The cosmic dance of the orbits is the music of the ice ages.

---

*I've drawn on it for much of the historical information in this chapter.

10. CYCLES OF DROUGHT

In 1878, scientist-explorer John Wesley Powell issued a strong warning against too-rapid expansion of population and agriculture onto the semiarid lands of the American West. The public, he realized, had little conception of how difficult the conditions for farming were on the western plains, west of the 100th meridian. Rainfall was marginal and unpredictable. "Many droughts will occur and many seasons in long series will be utter failures," Powell wrote in his typically forthright style in his official report for the federal government, *Lands of the Arid Region.* He repeated that forecast many times in the ensuing years.

Powell was already famous for his explorations of the Colorado River, his studies of Indian culture in the Southwest, and his insights into the natural history of the West. Geologist, naturalist, explorer, Powell was a man of intelligence, wisdom, and boundless energy. His one-armed condition (he had lost his right arm in the Battle of Shiloh) was less a handicap than an added element of glamour. Powell's books, articles, and lectures on his excursions in the Grand Canyon and the plateau country of the West were popular, and he was respected for his knowledge, courage, and outspokenness.

Despite all that, his warnings about cycles of drought fell on deaf ears. The 1870s had brought a dramatic improvement in climatic conditions in Kansas and other western states. Rainfall measured at army posts had increased up to 20 percent over the past twenty years and the prairie grass was lush and green.

Among many would-be farmers, there was an almost mystical belief, born out of these observations and a strong dose of wishful thinking, that somehow the very act of introducing farming to the semiarid lands stimulated a favorable change in climate. The cry was "Rain follows the plow."

Nonsense, Powell warned. "Usually such changes go in cycles, and the opposite or compensating change may reasonably be anticipated ... we shall have to expect a speedy return to extreme aridity, in which case a large portion of the agricultural industries of these now growing up would be destroyed."

Powell didn't have the historical information to know when the next such drought in the cycle would occur. He was certain, however, that it would. Throughout the 1880s he continued his pleas for less naïveté and more realism about the climate. In a famous speech in Bismarck, North Dakota, in August 1889, Powell, by now director of the U.S. Geological Survey, which he had helped create, once again took up the call.

"Years will come of abundance and years will come of disaster ..., and the thing to do is look the question squarely in the face." He called his listeners' view that the climate had changed permanently for the better a "delusion." The applause was polite and restrained. That wasn't what they had wanted to hear.

One year later, in the summer of 1890, the Great Plains states were plunged into disastrous drought. Across the newly settled lands of the West farmers were ruined. It had been twelve years since Powell's report had been issued forecasting the cyclical occurrence of such catastrophic events.

What Powell could not have known was the amazing regularity with which the cycles of drought would strike. We now know that the High Plains were hit by severe droughts in 1815–18, 1842–47, 1866–69, 1890–92, 1912, 1934, 1953, and again in 1976–77. Every twenty to twenty-two years drought visits its wrath upon the American West.

As I mentioned in chapter 8, by the time of the tragic Dust Bowl days of the 1930s, the roughly twenty-two-year periodicity of droughts had been noticed. The cycle became almost part of the American folklore. The recurrence of drought in the 1950s did nothing to dispel the idea. Throughout the late 1950s, the 1960s, and the early 1970s, popular and scientific periodicals alike were filled with predictions that the West would next be struck by drought in the mid-1970s.

I covered the annual meeting of the American Association for the Advancement of Science in February 1977 in Denver. The drought had then hit. The governors of eleven western states had just prepared a special drought task force. The 14,000-foot peaks along the Front Range of the Rockies, normally wearing a flashy crown of glistening white snow throughout the winter, were bare and brown. It was an ominous sight. Water is the critical element to the life and economy of the semiarid West, as Powell so long ago realized, and the snowpack created by heavy mountain snowfalls in the winter is the primary source of water for irrigation, industry, and personal use throughout the rest of the year. Five consecutive days of sessions on arid lands were held at that AAAS meeting, and the kickoff session, on American droughts, brought crowds overflowing out into the hallways. After a two-decade interlude, once again drought was on everyone's mind. The long-anticipated mid-1970s drought had arrived, just about on schedule.

Charles Greeley Abbot and other workers earlier in the century had pointed out that the drought cycles seemed to come at the same frequency as every other sunspot cycle. Then George Ellery Hale with his magnetograph had shown that the *magnetic* cycle of the sun was actually twenty-two years. The magnetic patterns of sunspots reversed themselves with each eleven-year sunspot cycle. So it took two sunspot cycles, or twenty-two years, for the pattern of magnetic polarity on the sun to repeat itself. In this sense, the fundamental cycle of the sun is twenty-two years. This reinforced the impression that something happening on the sun might indeed be influencing something on earth that led to droughts every twenty-two years in the western United States.

In chapter 8, I mentioned that Walter Orr Roberts became interested in the possible correlation between droughts and the twenty-two-year Hale double sunspot cycle sometime after his westward move from Harvard to Colorado. He was always careful to maintain that the apparent correlation might be only coincidence. Never-

HIGH PLAINS DROUGHTS

High Plains droughts and the twenty-two-year Hale sunspot cycle. Roughly every 20 to 22 years since 1818, and even before, the High Plains have experienced a serious drought. (Walter Orr Roberts)

theless, as he has said many times, "I have always considered the correlation very provocative."

He likes to show the association visually. He charts the sunspot cycles alternatingly. The peak of one goes upward on a graph, the peak of the next is plotted downward on the graph. This emphasizes the double sunspot cycle. He then draws in vertical lines representing the times of the eight High Plains droughts since 1800. The dates of the droughts correspond fairly well to every other solar minimum. Roberts calls the association "striking."

Yet could it all be coincidence? Many climatologists and solar scientists found the association intriguing. What was needed was some way to put it all on a much firmer, quantitative basis, and to extend the drought record further back in time.

And that brings us to the work begun in 1976 by Charles W. Stockton and David Meko of the University of Arizona Laboratory of Tree-Ring Research, drawing upon the legacy of A. E. Douglass. It was Douglass's interest in a sun-weather connection, you'll recall from chapter 8, that led him to studies of tree rings. Since his creation of the Laboratory for Tree-Ring Research in 1937, it had become a world-famous center for dendrochronology (establishing dates of events from the tree-ring patterns of wood associated with them). It had helped revolutionize first archaeology, then climatology.

The laboratory had an extensive collection of tree-ring climate indices. Harold Fritts and others had developed sophisticated ways to analyze the tree-ring climate record. They can even produce climate maps for given seasons in the past like the monthly weather maps the National Weather Service prepares now. No longer were climatologists limited to the sparse records of settlers and scattered military outposts for a view of nineteenth-century climate in the American West. And the beauty of it all is that excellent tree-ring climate records went back to A.D. 1600, before any European settlement of what is now the United States, except for the incursions of the Spanish up into the Southwest. The tree rings were a climate calendar, recorded independently of any human presence.

Stockton and Meko began a program to reconstruct drought patterns west of the Mississippi over the past three or four centuries. They carefully chose indices from tree sites especially sensitive to local climatic variations from the Mississippi River to the West Coast and from Mexico to Canada. They then established a series of indexes of extent of the geographical area over which drought existed for the entire West for each year back to 1600. This was a key. Some previous studies of drought *severity* didn't manage to show a clear connection between droughts and the double sunspot cycle. The technique Stockton and Meko developed gave a much broader overview, emphasizing the amount of land area affected by drought each year. This large-scale record of drought, they found, matches the double sunspot cycle remarkably.

Even to the eye, their charts of the chronology of drought areas show regular periodicity, at about twenty-two-year intervals, all the way back to 1700 and before. Some of the drought periods are more extensive than others. The Dust Bowl days of the 1930s affected a larger area of the United States than any other since 1700. That's probably small comfort to those who lived through it. The closest drought to it in geographical extent is one that peaked in the mid- to late 1730s. Other droughts affecting large areas came along in the late 1750s, the early 1780s, early 1800s, early 1820s, mid-1840s, early 1860s, and so on into the modern historical times.

Stockton and Meko were soon joined by the head of the climatology section of the National Weather Service, J. Murray Mitchell. Mitchell is one of the nation's most respected climatologists. A tall, slender man with glasses and a quietly self-confident manner, Mitchell is known as a cautious scientist, not given to voicing casual assertions. He came to the task of analyzing the Stockton and Meko data with a record as a strong critic of many proposed sun-weather connections.

During a visiting appointment in the advanced study program at the National Center for Atmospheric Research in Boulder, Mitchell subjected the data to a seemingly endless series of sophisticated tests. The first concerned whether the maximum area subject to drought really did peak at roughly twenty-two-year intervals, as it appeared to. It did. The studies showed that by far the predominant cycle was one of about twenty-two years. The twenty-two-year periodicity was real at what scientists call "the 99 percent confidence level." The observed twenty-two-year rhythm was not likely to be a chance occurrence.

The next task was to compare the drought cycle with the twenty-two-year double sunspot cycle and see if they were locked in phase. Again all sorts of tests were carried out. The verdict: "Strong evidence of a systematic phase locking between the 22-year rhythm of large scale drought in the western U.S. and the Hale sunspot cycle since 1700 A.D." The relationship was this: Drought was most favored to reach its maximum extent within the first two or three years following alternate minima in the eleven-year sunspot cycle. This result too was found to be highly statistically significant. It was strong evidence that the drought and sunspot rhythms are indeed connected.

Mitchell and his colleagues then also looked at whether there was any additional connection. Was there any relationship between long-term changes in the heights of these cycles? In other words, during a period in which the sunspot cycle was particularly strong or weak did the drought cycle show any similar effect? They found a suggestion of such an effect. Specifically, epochs of especially large drought cycles were followed by especially large sunspot cycles. The clearest example was with the Dust Bowl drought of the 1930s, the largest drought in two centuries. It was followed within twenty years by the highest sunspot numbers ever recorded. Other periods showed similar effects. The drought rhythm was quite large going into the Maunder sunspot minimum around A.D. 1645 but decreased to a very low value near the end of the Maunder minimum around 1700. Both the drought and sunspot rhythms were also low around A.D. 1900. These apparent associations were not quite so statistically certain as the others but were nevertheless intriguing.

The important point is that Mitchell, Stockton, and Meko had finally demonstrated rather convincingly that large-scale droughts do indeed occur at twenty-two-year intervals and that this cycle fits the twenty-two-year double sunspot cycle like a key in a lock.

This work had been followed closely by all scientists interested in sun-weather connections. As each step in the analysis was completed, Mitchell and his coworkers seemed a little more certain. They finally presented their results at a meeting on solar-terrestrial influences on sun and climate held at Ohio State University in August 1978, the proceedings of which were published the following year.

Their three kinds of evidence, they said, strongly support the idea that the extent of drought in the western two thirds of the United States varies in a pulselike manner on the same twenty-two-year frequency of the Hale magnetic cycle of the sun. The results have high enough statistical significance, they said, to justify "an earnest inquiry" into possible cause-and-effect mechanisms.

They admitted being a little shaken by the results. "As hitherto avowed agnostics with regard to many previous claims of sun/climate relationships, we find ourselves somewhat unnerved by our own data."

They said they would take comfort in any corroboration—or even refutation—of their conclusions through any independent investigations by other scientists. Most scientists now agree that the connection has been put on fairly solid footing.

What does it all mean? Mitchell, Stockton, and Meko summarize: "Our results would clearly seem to imply a role of solar magnetic activity in giving rise to widespread drought in the western United States." How all this could come about is now the question.

The role need not be direct. Solar control probably is not necessarily the prime mover of drought or even of climatic aberrations that lead to drought. "Rather," say Mitchell and colleagues, "we prefer to think that the solar control is in the nature of a modulating mechanism, that alternately favors or discourages the spread of drought at times when terrestrial climatic developments unrelated to solar events are primed to erupt into a drought situation." The solar connection, in other words, just gives the final nudge.

Some general forecasts are made possible by this confirmation. It would hardly be going too far out on a limb, for instance, to suggest that the next period of widespread drought across the western United States will come in the mid- to late 1990s. But that's still about all you can say. Certainly it enables no climate "prediction" on any finer scale. As Mitchell's group points out, a wet year can well arrive at a time when the sun "says" it should be a drought year. And a major drought can develop when the sun "says" there should be no drought.

Take the severe heat wave and drought that struck Texas and other southern and midwestern states in the summer of 1980, for instance. Elderly and sick people without access to air conditioning suffered miserably. At least 1,265 persons lost their lives as a direct result of the heat. Crops were ruined. Millions of poultry died. Highways buckled. Cars broke down. A federal study estimated the economic cost of the heat wave at $20 billion. Yet it is not at all clear how that event fits into this picture. The low point of the last Hale sunspot cycle was in June 1976. By that standard the 1980 heat wave and the continuing drought through the 1980–81 winter snow season came a year too late to be legitimately associated with the two-to-three-year lag the Mitchell-Stockton study finds is the usual interval between Hale sunspot minimum and drought. Could the 1980–81 drought be considered just a little late but still part of the expected drought interval that began with the 1976–77 western drought? (Walter Orr Roberts thought that might be the case. "It is right at the end of the Mitchell-Stockton 'period of enhanced drought risk,' " he said in February 1981. "It's right on the border line! Again ambiguity!") Or is it just totally out of phase, thus not fitting into the picture at all? Also, intense as it was, it is not clear the heat wave was widespread or long enough in duration to show up eventually as an important high point on a long-term record of drought-area index. Remember, the correlation Mitchell and Stockton find with the Hale sunspot cycle is not *severity* of drought but geographical *extent* of drought. I point out all this merely to show that these associations are not simple. The new work is best at associating large-scale long-term drought patterns with the solar cycle, not small-scale short-term ones.

What this landmark study *does* show, in the concluding words of the investigators, is: "The *risk* of widespread drought somewhere west of the Mississippi River—exact location unspecified—is appreciably higher in the years following a Hale sunspot minimum than it is at other times during the Hale cycle."

What's so special about the western plains of America? Why

does an apparently sun-related drought cycle of twenty-two years show up there and, so far, nowhere else in the world? Isn't that suspect? Well, for one thing, the semiarid western United States is especially amenable to tree-ring climate studies. The tree-ring climate record is most pronounced in areas of marginal climate—where a good or bad year of weather can stimulate a striking change in the growth of a tree. In areas of more abundant rainfall, tree rings don't yield such a clear record. There, many other factors, such as density of trees and other nonclimate local conditions, are more critical elements.

But there is indeed something geomorphically unique about the western plains. Walter Orr Roberts points out that the Rocky Mountain chain is the largest single north-south mountain barrier to the mid-latitude westerly winds that exists anywhere on earth. Perhaps, he suggests, it may be responsible for a generally permanent ripple in the general atmospheric circulation that results in drought when the westerlies are strong and persistent and impede the northward movement of moisture from the Gulf of Mexico. If persistent westerlies can be influenced in some cyclic way by changes in solar magnetic activity, that might in turn explain the pronounced solar-linked drought cycles of the western plains. This, of course, is just a speculation.

There's also another kind of scientific evidence associating climate fluctuations with the twenty-two-year Hale solar cycle. Curiously, it too comes from trees. (Will climatologists and solar scientists ever be able to repay their profound debt to the world's trees?) In this case, the telling clue is not the pattern of widths of annual growth rings but the chemical composition of the wood in those rings.

About 14 of every 100,000 hydrogen atoms in nature are actually a heavier form of hydrogen called deuterium. A small proportion of the molecules of water vapor in the air therefore contain not regular hydrogen atoms but deuterium. Molecules of water vapor made of two deuterium atoms and an oxygen atom are slightly heavier (by one ninth) than the normal water-vapor molecules made up of two hydrogen atoms and an oxygen atom. When a mass of moist air containing both types cools, the heavier water molecules—those made up of deuterium instead of hydrogen—naturally precipitate out first. So as an air mass cools it becomes progressively depleted in deuterium. This means in warm areas, such as the southeastern United States, the rainfall and surface water contain fairly high proportions of deuterium. In cold areas, such as Alaska, they are depleted in deuterium. The hydrogen and deuterium atoms in the water eventually get incorporated into the cellulose in the wood of trees. The ratio of the two isotopes present in the water available to the trees during their growth periods becomes "locked into" the wood.

The isotopic ratios in the wood of tree rings are thus a kind of fossil thermometer of temperatures at the time that ring was growing. By analyzing these changes over long periods of time as revealed in the tree rings of especially long lived trees such as the bristlecone pine it is possible to get an idea of climatic variations over long periods of time. It is also possible to analyze these changes for the presence of climatic cycles.

Geochemists Samuel Epstein and Crayton Yapp at Caltech have analyzed a thousand-year record of hydrogen in cellulose of bristlecone pine trees from the White Mountains in California. They wanted to see if there was any clear interval of climate fluctuation. There was. A strong twenty-two-year cycle showed up. It was there whether they analyzed the data ten rings at a time or five rings at a time. The twenty-two-year climate cycle was apparent for the past thousand years of tree growth. The cycle was strongest during the past two centuries. Why is not exactly clear. What is more interesting is that during the period of the Maunder sunspot minimum, the twenty-two-year cycle disappears from the chemical record in the wood. There are fluctuations in the hydrogen-deuterium ratios during the Maunder minimum; they just don't show any twenty-two-year cycle.

If all this is indeed related to events on the sun, it once again emphasizes that the critical factor is the twenty-two-year Hale *magnetic* cycle, the so-called double sunspot cycle, instead of the visual sunspot cycle. Epstein believes the relationship between the twenty-two-year cycle in the tree isotope data and the solar cycle is probably real. The trees seem to be relating real information about solar-related climate fluctuations.

Some caution is necessary. Although certain other trees show the twenty-two-year cycle as well—a cedar tree in Sequoia National Park reveals an unusually excellent one—some other trees elsewhere don't. Pine trees from Oregon and Scotland show no evidence of it. Many local conditions may influence whether a tree does or does not record a cycle. Much more work has to be done.

For now, however, solar scientists are pleased with the

Epstein results. They seem to be a confirmation of a twenty-two-year solar effect on climate. And the possibilities for further study are tantalizing. Cellulose has been found in plant material that grew thousands and even millions of years ago. The ancient remains of once-proud trees may play a pivotal role in eventually establishing a clear sun-climate connection.

What could that connection be? We don't really know. A respected Princeton University theoretical physicist, Robert H. Dicke, sees in the deuterium-hydrogen data a big clue supporting an intriguing theory he has about what's going on inside the sun.

Dicke first of all notices that the visual sunspot cycle, although it averages 11 years, is really not all that regular. Sometimes it's a year or so longer, sometimes shorter. The intervals from one maximum to another have been as short as 7.3 years and as long as 17.1. What's important is that the sun seems to correct itself from getting out of phase for too long a time. For instance, if it has, say, two short cycles in a row it will follow them with an especially long cycle. This brings the timing back to the 11-year long-term average.

It's as though a clock running slow had the ability to sense that fact and correct itself. Every few hours it could detect whether the minute hand is back at the twelve position; if not, it gives it a little corrective advance.

This is what the sun seems to do. It's almost as though there is some sort of clock inside the sun keeping track of the timing. "Is there," Dicke asks, "an internal clock to improve the sun's memory?"

What has all this to do with the twenty-two-year cycle in the Epstein-Yapp tree isotope record? Well, Dicke notes, the twenty-two-year periods seem to come very regularly, never missing the beat. They don't seem to show the irregularities we see in the visual manifestation of the sunspot cycle. Their rhythm stays sharply tuned. And remember, the twenty-two-year magnetic (double sunspot) cycle is apparently the more fundamental rhythm of the sun, since magnetic changes drive the other features. If the twenty-two-year cycle is indeed a sun-driven feature, its regularity in the tree isotope data seems to imply that it has some direct connection with what Dicke refers to as a "precisely tuned internal oscillator" in the sun.

In this view the sunspots, faculae, flares, prominences, or solar wind have nothing fundamental to do with it. They are, as Dicke puts it, so much "minor surface fluff," just by-products of more vigorous processes in the solar interior. They all get temporarily out of phase with the regular rhythm. But the deuterium-hydrogen fluctuations in the trees don't. They are tied, he suggests, to the sun's internal clock.

This deeply buried clock, whatever it may be, would probably lie at the base of the solar convection zone. It would be a kind of magnetic-fluid oscillator that both controls the magnetic cycle of the sun and modulates some fundamental feature of the solar output such as, perhaps, its luminosity, at twenty-two-year intervals. If so, the kinds of precise long-range measurements of solar luminosity I described at the end of chapter 7 take on even more importance as a means of proving or disproving Dicke's hypotheses.

Dicke's idea is admittedly highly speculative. And it is based, really, on only one set of data, the precise twenty-two-year variation seen in Epstein and Yapp's measurements of deuterium and hydrogen in tree rings. It will be interesting to see whether future work lends more credibility to the idea. It is an intriguing thought that the sun may have a precisely tuned internal clock whose timing sends impulses that modulate detectable changes in our planet's climate.

In the meantime, the Mitchell-Stockton-Meko work confirming a twenty-two-year drought cycle locked in phase with the Hale solar cycle stands as the most solidly based evidence of a sun-climate connection. Epstein and Yapp's isotopic tree-ring data lend still further support.

Together, if they stand up to further tests, they point to a solar forcing factor operating on climate over a long period of time. The challenge next facing scientists interested in the sun-weather connection, in the view of Walter Orr Roberts, is then to relate these results to evidence of sun-weather connections on the scale of weeks and days. To that subject we turn next.

11. SECTORS, STREAMS, AND STORMS

We live in the flow of the sun's star streams. Not until the space age, however, did we gain the ability to transport our senses out into space between the sun and the earth and detect the structure of that flow. That new vantage point was exhilarating. All those little spacecraft scooting about the inner solar system for the first time revealed something that had never been apparent before.

They not only confirmed the existence of the solar wind—something that solar physicist Eugene Parker had predicted theoretically—but also showed that it was divided into sectors that rotate with the sun. The solar wind, as I noted in chapter 2, is gaseous material of the sun spewing out into space. It races out across the solar system at anywhere from 300 to 700 or more kilometers per second, or 650 thousand to 1.5 million miles per hour. At these speeds it reaches earth in roughly four and a half days.

If the solar wind were visible and if we had a view far above the solar system, the appearance would be much as though the sun were a rotating water sprinkler. The solar particles trace a graceful spiral trajectory.

Using data sent back by spacecraft, American geophysicists John M. Wilcox and Norman Ness were able to show in 1965 that this

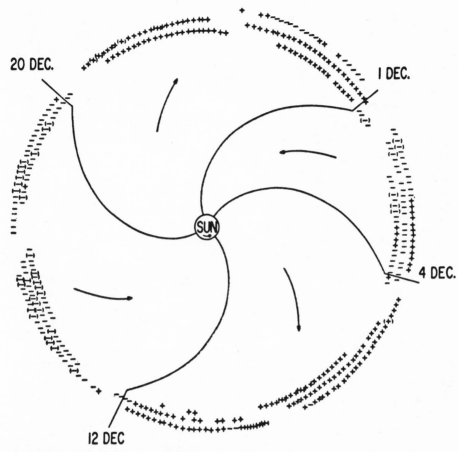

The pinwheel-shaped solar sector structure of the interplanetary magnetic field as first detected by satellite in the 1960s. Adjacent sectors have magnetic fields of opposite polarity (arrows) streaming out from the sun. This entire structure rotates with the sun once each 27 days. (Wilcox and Ness)

rotating flow was divided into roughly four sectors. Again looking down from our hypothetical vantage point above the solar system, the four quadrants would look something like the vanes of a pinwheel or the petals of a huge flower. Or, to return to the water-sprinkler analogy, the solar sprinkler spews out four equally spaced broad streams of particles.

Now, the solar wind carries more than just particles. It also stretches out across space with it the sun's magnetic field. Since this magnetic field originating in the sun extends out across the space between the planets, we call it the "interplanetary magnetic field." What really distinguishes adjacent "sectors" of the solar wind structure is that they are of opposite magnetic polarity, like the opposite ends of a magnet. One sector is positive, the next negative, and so on. Solar scientists prefer to use slightly different terminology for the polarity of the solar-wind structures. They refer to one sector as having a magnetic field directed away from the sun (positive), the next as having a magnetic field directed toward the sun (negative). This entire structure rotates with, and is carried away from, the sun, just as the pattern of water from a sprinkler rotates with the sprinkler head.

That's the solar sector structure as diagrammed on paper—a flat, two-dimensional view. But in recent years scientists have been able to infer something of the three-dimensional shape of this structure. This invisible structure in space has been given some body.

The illustration, courtesy of Wilcox, helps show it. You can see that these regions of opposite magnetic polarities in interplanetary

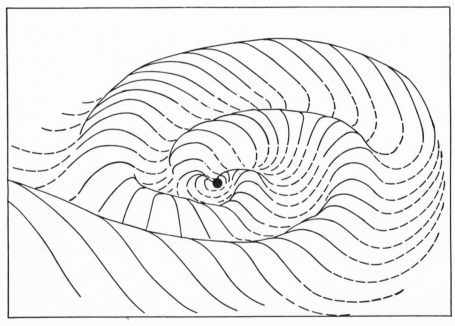

In a three-dimensional view, the solar sector structure becomes a warped current sheet in the inner solar system, rotating with the sun. The polarity of the interplanetary magnetic field originating with the sun is at present positive "above" the warped surface of this current sheet, negative "below" it. (Svalgaard and Wilcox)

space are in reality separated by the warping upward and downward of what solar scientists now call the warped heliospheric current sheet. In this three-dimensional view, this current sheet divides the interplanetary magnetic field into two regions with oppositely directed lines of magnetic polarity. The region above the warped surface of the current sheet has a magnetic polarity in one direction. The region below its surface has a magnetic polarity in the opposite direction. This results in a four-sector structure brought about by the warped shape of the surface of the current sheet.

As I mentioned, this entire structure rotates with the sun every twenty-seven days. The earth is never more than a slight amount above or below the plane of the sun's equator. This means the warps in the current sheet are normally large enough that as the sun rotates, the surface of the current sheet, the so-called sector boundary, over-takes the earth roughly four times every twenty-seven days. The polarity of the interplanetary magnetic field will thus change four times every twenty-seven days, or roughly once every seven days.

In this new three-dimensional view there is, in reality, only one "sector boundary" in space—the warped current sheet whose ruffled surface separates positive and negative magnetic fields.

The size of a particular sector may change somewhat from one rotation to the next. And sometimes the warped current sheet flattens so much that only two sectors sweep by earth every solar rotation. So the twenty-seven-day periodicity is not always strictly observed. But it is a good approximation of the long-term case.

How does such an elegant swirling structure originate on the sun? The light from the sun has information about the sun's magnetic fields in it. Every day magnetographs at observatories such as the Stanford Solar Observatory in California analyze that light and make charts of the sun's magnetic field. The long-term average reveals a basic pattern. There are large areas of positive and large areas of negative magnetic polarity on the sun. They tend to alternate in about four large regions around the sun. The line separating them winds its way up, down, and across the solar surface in a fairly regular pattern.

The pattern is very similar to that on a baseball. The surface of a baseball is made of two pieces of leather, joined along one continuous, curving seam. Now imagine the same-shaped pattern on the sun, each piece having opposite magnetic polarity. On the sun these

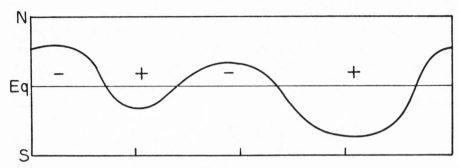

Like the hide of a baseball, the surface of the sun (here flattened to two dimensions) is divided into two large interlocked sections of opposite magnetic polarity. This helps explain the sector structure of the warped current sheet. The neutral line, or "seam," becomes the surface of the current sheet when extended outward into space. (Svalgaard and Wilcox)

regions extend about half the way toward the pole from the equator, although that amount can vary considerably.

The magnetic fields of these regions are carried by the solar wind hundreds of millions of kilometers out into interplanetary space. The areas of positive and negative polarity are separated by an undulating surface that is the projection into space of the magnetically neutral line on the sun (the seams of the baseball).

The magnetic sectors on the sun can last a long time— decades or longer. And as long as they are there, the persistent warp they cause in the current sheet in space is maintained, all the time rotating with the sun like a slowly turning carousel.

On the sun, the middle sections of these large magnetic regions, like the polar caps, often—although not always—contain coronal holes. As I mentioned in chapter 2, coronal holes are low-density areas in the outer solar atmosphere where the magnetic lines of force are open and especially high speed streams of solar wind can escape outward from the sun. The fundamental influence, however, may be the persistent warp in the current sheet caused by the magnetic sector. The coronal holes may just be accidental side effects of this large-scale magnetic configuration, according to Wilcox and colleague Leif Svalgaard.

What sort of physical reality is there to this solar sector structure in space, apart from the alternating magnetic polarities? Around 1977 and 1978 it became apparent to solar scientists that the speed of the solar wind is always low (down to about 300 kilometers per second, or 700,000 miles per hour) at any sector boundary. At some distance away from the neutral line, the solar wind speed is much greater (around 750 kilometers per second, or 1.7 million miles per hour). As Svalgaard and Wilcox point out, this means that if the warp in the current sheet is wide enough and deep enough, we will see a high-speed solar wind stream emanating from the solar magnetic sector causing the warp. So the earth is alternately buffeted by slow and fast solar winds during each sector passage, which lasts about seven days.

What has all this to do with the weather? Well, let's go back a bit now. Back in 1965 when the solar sector structure was first reported, scientists also immediately noticed that geomagnetic activity—fluctuations in earth's magnetic field—increased after each solar sector boundary sweeps past earth. Indexes of geomagnetic activity show sharp peaks about two days after transit of a solar boundary, then a quick tailing off over the next five days. That effect has been confirmed many times over the past decade and a half.

So, to summarize, we know that the time just before a boundary transit the speed and density of the solar wind, the magnitude of the interplanetary magnetic field, and geomagnetic activity on earth all tend to be at a minimum. All of these quantities begin rising sharply as the boundary sweeps by earth and all reach a maximum about two days later.

You'll recall from chapter 8 that back in the 1950s Walter Orr Roberts and his colleagues had noticed an apparent sun-weather effect. The record of persistence of weather patterns rose and fell in a consistent way according to how many days had elapsed from geomagnetic storms. And a little later they showed that winter storms forming in the Gulf of Alaska were larger if they had been preceded by a geomagnetic storm than if they hadn't been. These correlations persuaded Roberts that the sun's activity did have some demonstrable effects on weather patterns.

The connections seemed to have something to do with times of sun-caused geomagnetic fluctuations, but what really was happening? The first work had been done before Sputnik, before we had any electronic "eyes" out in space to probe the interplanetary medium and detect the sector structure of the solar wind and the magnetic fields it carries with it. But now the sector structure had been identified and geomagnetic effects were seen to rise and fall in harmony with the transits of sector boundaries roughly every seven days. Might this be an essential key?

Roberts realized any association with weather patterns would have to be put on a firm statistical footing, free of subjective bias. A longtime colleague and collaborator of his, Roger H. Olson, a meteorologist then doing solar research in Boulder, hit on the idea of a new and much more accurate index. Derived by computer from twice-daily official weather maps, it measures the total area of the northern hemisphere where there are extremely strong cyclonic storms. Cyclonic storms are those circulating around a low-pressure area, and that includes almost all the large storm systems affecting the hemisphere. Olson and Roberts developed this measurement, called the vorticity area index. When the number is high, the area covered by strong cyclonic storms is large. When it is lower, less area of the northern hemisphere is covered by strong storms. John Wilcox, Philip Scherrer, and Leif Svalgaard at Stanford set to work comparing fluctuations in the vorticity index with times when solar sector boundaries (the warped current sheet) cross the earth.

In a series of papers beginning in 1973 and continuing right up into the 1980s, they have time and again reported a clear pattern of change. About one day after a solar sector boundary crosses the earth the vorticity area index takes a sharp dip. The area affected by large low-pressure storm systems drops. Then just as quickly the indexed area climbs back to normal levels. The amount of change is about 10 percent.

The pattern is quite pronounced and consistent. Something seems to happen in the lower 10 kilometers (6 miles) of the atmosphere where all storms occur (the troposphere) to sharply reduce for a day or so the strength and extent of large circular storm systems.

I have seen the graphs charting these changes many times, and I still find the consistency and depth of the effect remarkable. So does Wilcox. "I get a feeling," he says, "that after boundary transits something may have grabbed the tropospheric circulation in the entire northern hemisphere."

The pattern stands up no matter how they are examined. It is

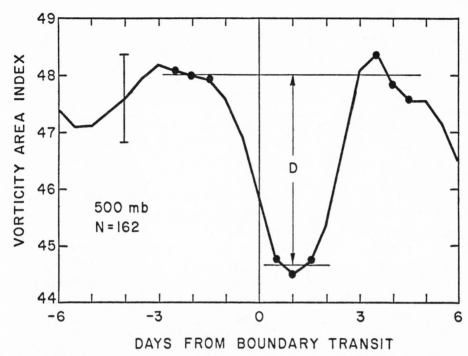

Repeated studies show that the amount of vorticity (storminess) in the northern hemisphere during the winter takes a pronounced dip in the day or two after the boundary of a solar sector sweeps past the earth. (Wilcox and Scherrer)

seen in the data for the years 1964–66, 1967–70, and 1971–73. It is seen in different bands of latitude from 20°N to 90°N. It is seen whether the sector change transiting earth is going from positive to negative or negative to positive. And it is seen in both the first half and the second half of winter.

The effect is prominent only during the five winter months, November to March. This has bothered many critics. But Wilcox points out that in winter the temperature differences between the earth's polar regions and equator are at a maximum. This intensifies the circulation of the atmosphere and produces the largest stresses on it. Perhaps the atmosphere is more susceptible to a solar influence then.

One of the strongest early critics of these results was Colin Hines, an atmospheric physicist at the University of Toronto. Hines didn't believe the claimed effect. The statistical significance of the correlation had been tested, he said, "in only the most rudimentary of fashions" and there was no "quantitative evaluation." If the claim were true, he realized, its implications would be far-reaching. He set out to disprove it. As he says, he and a colleague "sought several means of undermining the credibility of this claim."

Wilcox and Svalgaard sent him their latest data, and Hines went to work on it. He soon was astonished. Using all the available data and subjecting it to all manner of new tests, Hines found that the correlation stood up handsomely. His graph of the vorticity index showed the same pronounced dip one day after solar sector passage as Wilcox's had. They were in fact virtually identical.

Hines had found no way to reduce the statistical reliability of the correlation below the "95 percent confidence level," meaning it was highly likely to be a real relationship.

"The claimed effect has been challenged in some depth and, as it turns out, has survived the challenge," Hines and colleague Itamar Halevy stated in their detailed report in 1977 in the *Journal of the Atmospheric Sciences.* "We take the existence of a meaningful correlation to be established by these new results."

It was a classic example of the workings of the scientific method. An intriguing and dramatic claim had been challenged by knowledgeable skeptics who set out to subject it to the most rigorous independent tests they could concoct. That overcomes any reservations that the original investigators' own interest in the outcome of the results might somehow have unconsciously influenced their analysis or conclusions.

Hines and Halevy's original skepticism had been thoroughly allayed by their own analysis of the data. "The classical scientific method—that of analysis, prediction and test—then demands that we accept the correlation as having been established as physically meaningful, and we have invited our readers to do likewise."

They suggested that the solar influence responsible for the correlation, whatever it may be, acts somehow to slightly alter the timing of naturally occurring events in the atmosphere. In other words, the solar influence doesn't in itself have to produce a major effect in the atmosphere; it only gives a little nudge that knocks effects in progress temporarily off their stride, delaying them by as much as a day.

Roberts calls the Hines and Halevy analysis "the most careful and the most elegant" of all attempts to test the correlation Wilcox, he, and the others had been reporting. Their tests of significance "were far more sophisticated than ours," he says. Needless to say, he and Wilcox were pleased.

"This finding," Roberts told a sun-weather conference in Ohio in 1978, "establishes a relationship between solar activity and the weather that is very hard to disbelieve, as implausible as it may seem."

Two Cornell University atmospheric scientists, Miguel Larsen and Michael Kelley, have reported a completely independent repetition and extension of Wilcox's findings using data from different time periods and sources. "We have found the same decrease in vorticity area index one day following sector boundary crossings," they said.

Not everyone agrees. Physicist Gareth Williams, pointing to an apparent weakening of the effect in the few years after 1973, insists it is all due to "statistical chance." Some other scientists have also expressed reservations. Wilcox replies that the effect is still clearly there and that the shallower vorticity dip may be due to a reduction in large-scale atmospheric circulation during those years. And the arguments go on.

There was another possible effect to search for as well. So far all the correlations had dealt only with the *time* of solar sector transit, not the magnetic *polarity* of the sector. Does the storm area index show any difference depending on whether the solar sector sweeping past earth is magnetically positive or negative?

A young student at Harvard, P. B. Duffy, then still an undergraduate, spent a summer working with Wilcox at Stanford examining this question. With the aid of Olson and Roberts he pored over data on individual low-pressure storm systems in the Gulf of Alaska, plotting their specific vorticity area indexes daily for a full twelve days as they moved eastward.

It turned out that the low-pressure troughs tended to be larger if they were first seen in the Gulf of Alaska while the solar sector sweeping earth had a positive (away-from-the-sun) polarity. If the solar sector had a negative (toward-the-sun) polarity, the storms tended to be smaller than average. As Roberts points out, this result indicates that "there is a specific geographical region, the Gulf of Alaska, where the sun-weather effect is especially strong." Here was another apparent solar-related effect on storms in the Gulf of Alaska to go along with the one Roberts had first tentatively identified back in the 1950s (chapter 9). Why would it be strong in the Gulf of Alaska? Roberts suggests some possibilities. The contrasts in temperature between the gulf waters and land are strong during the winter. The gulf is at high latitudes where sun-caused geomagnetic activity may have a somewhat stronger than normal effect. And extremely strong storm systems are generated in the gulf in winter. At a workshop in the sun-weather question at Stanford in 1980, Jerome Namias, a Scripps Institution of Oceanography research meteorologist, characterized the development of low-pressure storm systems in the Gulf of Alaska during winter as "very explosive." The gulf is an unusual area where a small perturbation can produce a large effect. This is especially true when the winds blow from the north, bringing cold air from northern Alaska down over the warmer gulf waters. Perhaps one or more of these characteristics have something to do with the apparent solar effect. At any rate, the Gulf of Alaska is the source of many of the storms that march across North America at regular intervals in the winter.

And that brings us to an extremely interesting second aspect to the Larsen-Kelley study I mentioned earlier. The vorticity area index is, after all, a measure of the extent and intensity of large storm systems. A temporary dip should have effects detectable in other ways. In fact, if our regular weather forecasting procedures don't include this effect in their forecasts—and they don't—you would think that the accuracy of the forecasts would also take a dip for the times the solar influence is apparently at work.

Larsen and Kelley looked at that possibility, and that is exactly what they found. For the previous three winters they compared the accuracy of the official computer-generated National Weather Service forecasts for North America with the times of solar sector boundary transits of earth. It wasn't a trivial task. The weather forecasts had to be converted into the lingo of the vorticity area index for comparison.

They found that the accuracy of the weather forecasts took a decided dip in the two days following the boundary transits. That fact was true of both the twelve-hour and the twenty-four-hour forecasts. The accuracy pattern looks hauntingly similar to the vorticity index pattern. Their results suggest that the usual forecast accuracy of 82 percent drops to about 68 percent directly after the sector boundary transit.

"This result is one of the first in which the sun-weather effect nas a potential practical significance," said Philip Scherrer in a recent review of the effects of solar variability on weather.

Walter Orr Roberts also considers this result extremely important. "Scientists at the National Weather Service and elsewhere struggle from year to year to get a few percent improvement in the quality forecasts," he points out. "This result suggests that the present weather service forecast model has a source or sink of energy missing from it that is related to solar activity. If we can figure out what the process actually is, we may have a new means of improving forecasts at the specific times when the solar effects appear to be operating."

It is a tantalizing possibility that we may already have nearly at hand a way to noticeably improve weather forecasting accuracy. Yet the data are still limited, and we as yet have no real way of understanding exactly how the solar influence, whatever it may be, is being transmitted to earth's weather machine. The meteorologists in charge of our nation's computer-generated forecasts can hardly be blamed for not adding to their forecast models an apparent sun-weather effect that cannot yet be physically explained. Yet I can't help wondering whether in a decade or two the weather forecasts you hear and see on TV predicting the next assault of a winter storm will be aided by routine inclusion in them of the solar sector factor, and be slightly more accurate because of it.

In reviewing the status of the quest for solar influences on earth's weather and climate in *Nature* a few years ago, UCLA atmospheric scientist George Siscoe listed the three strongest cases. Although he said it was still too early to have "unreserved confidence" in them, all three revealed "what seem to be well-resolved sun-weather signals." Their claimed correlations, he said, should now be granted the status "of reasonable to virtual certainty."

The three cases? The first was the seeming correlation of climate variations lasting a few centuries with Maunder-minimum-type excursions in solar activity. The second was association of drought cycles and the Hale twenty-two-year double sunspot cycle. The third was the effect I've been describing in this chapter on the amount of large-scale storminess a day after the passage of solar sector boundaries. I have devoted nearly an entire chapter to each of these three cases. A fourth near-certain case would undoubtedly be the association of ice-age cycles with the variations in the orbital geometry I described in chapter 9, but Siscoe was considering only cases where something about the sun itself was known to be varying.

I have deliberately emphasized these four cases because I wanted to consider only the *best* evidence of influences of the changing sun on earth's weather and climate. I've already shown, at the end of chapter 8, that the case for Maunder-minimum-type effects on earth's climate no longer looks quite so good as it did a few years ago, although the jury is still out. The drought-cycle association, however, and certainly the orbit–ice-age link, have held up very well. The apparent effect of sector boundary transits on short-term weather is not yet quite as firmly accepted as these others, in my view.

Roger Olson, who now works with Walter Orr Roberts at the Aspen Institute in Boulder, mentions a few more sun-weather relationships that "appear to be both valid and of physical importance."

One is an association between thunderstorms and events on the sun. Both European and U.S. data have shown that thunderstorms increase in frequency a few days after the occurrence of large solar flares near the central meridian of the sun (when they are directly facing earth). Olson was completing a study of this connection in early 1981 and said the correlation looked very good.

The second is the deepening of low-pressure troughs emanating from the Gulf of Alaska following geomagnetic storms, the work Roberts and Olson pioneered. It has enormous potential practical importance, and they have recently expanded their studies using the computer at the National Center for Atmospheric Research.

A third is the heating of the upper atmosphere (the stratosphere) by solar activity, the effect that prematurely doomed Skylab. The issue is not the stratosphere itself but how the effects of this heating might influence the troposphere, the lower part of our atmosphere, where all the weather takes place. Also of concern here are possible weather effects of sun-caused changes in the ozone layer.

A fourth is a possible eleven-year cycle in surface temperatures over North America.

There are still other possibilities as well. But they are all less certain. They need independent confirmation. Scientific works on the sun-weather question mention literally hundreds of possible associations. Few of them have anywhere near the status of acceptance of those I have been describing, and I see no reason to go into them now.

As I emphasized in chapter 8, to many scientists the accumulated evidence for sun-weather connections is persuasive. To a few others, it's all an illusion. In the recent words of A. Barrie Pittock, the strongest skeptic of all, in a 1980 review in *Nature*: "This is a highly controversial and—as I know from personal experience—an emotional subject. . . . I wish the philosophers of science would address themselves to this particular controversy because I suspect that in the literature on the solar weather/climate relationships they will find their best possible case study of objectivity versus subjectivity, and of the conscious or unconscious influence of vested interests and disciplinary prejudices." Most scientists familiar with the field fall somewhere in between the two extremes But even avid, well-informed proponents of sun-weather connections agree that despite the intriguing and even exciting positive evidence I have been discussing in the last few chapters, the case for clear sun-weather effects is far from proved, in the rigorous sense in which scientists refer to proof.

Olson, despite being an acknowledged proponent, calls the present situation a "crisis in sun-weather research." What is the problem? Just look at all the evidence I've been describing, let alone dozens of other claimed correlations. Isn't that enough? No. It wouldn't be enough even if all the correlations were statistically pure. Standards of accountability in science are far more strict than in other fields of human activity, such as politics. Claims really can be tested. Only when they meet the most stringent tests are they adopted. That is one of the things I find so refreshing and wondrous about science.

And what is wrong with all the abundant evidence of sun-weather correlations? Nothing. Except that scientists have yet to come up with any agreed-upon way to explain what might account for them. Olson says it well for all scientists in this field, critics and proponents alike: "The crisis in sun-weather research is caused by the fact that we have a great outpouring of new empirical results, without a corresponding increase in our understanding of what is happening physically to explain them."

Scientists need to understand exactly how variations in the sun and its emanations can tickle the earth's lower atmosphere to cause weather and climate change. Only then, despite all the dramatic new results I've been describing, will the logjam blocking further progress in sun-weather research be broken.

I take up the quest for that missing mechanism in the next chapter

12. SEARCH FOR A MECHANISM

Nothing makes a scientist more uncomfortable than a proclaimed effect he or she cannot explain. The unease is not just due to ego or frustrated curiosity. It stems from the inability to eliminate the disturbing possibility that the effect may not be real. If there is no satisfactory physical explanation for the apparent connection, how can we be sure coincidence is not the only operative force? How can we be certain we are not fooling ourselves?

Nowhere is this more true than with statistical correlations. And few statistical correlations are more subject to such gnawing doubts as those amassed in behalf of sun-weather connections. What physically is happening? What goes on between the sun and the earth's lower atmosphere to cause the supposed effects of changing solar emanations on air circulation and weather patterns I have described in the previous two chapters?

As Walter Orr Roberts said to me concerning his group's intriguing proposed sun-weather connection, "Until one can find a viable physical mechanism and verify it, no one will believe it's true."

The leading candidate for a mechanism to explain the sun-weather connection concerns the electrification of the atmosphere. The basic idea leads all the way back to the prominent Scottish physicist C. T. R. Wilson. In 1916, five years after his perfection of the

Wilson cloud chamber for tracking nuclear particles (which eventually brought him a Nobel Prize), Wilson proposed the existence of a global atmospheric electrical circuit.

The idea is this. The earth's atmosphere lies between two electrical conducting plates. The bottom plate is the solid surface of the earth, which readily conducts electrical current. The top plate is the ionosphere, the layer in the atmosphere beginning about 70 kilometers (40 miles) up, where significant quantities of the atoms have become ionized, converted into electrically charged particles. The ionization is caused by interactions of both cosmic rays from outside our solar system and ultraviolet radiation from the sun on the atoms in the atmosphere. The result is that this entire shell of the atmosphere can conduct electricity.

Wilson realized that there is a global electrical circuit, sometimes now called the Wilson circuit. Current flows upward to the ionosphere from thunderstorms and travels horizontally across the shell of the ionosphere. It then makes its way back down to earth through the

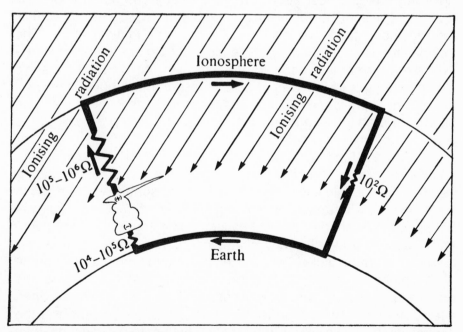

The atmospheric electrical global circuit. The thunderstorm depicted represents the global electric generator—the totality of all thunderstorms in progress on the planet. (Courtesy Ralph W. Markson)

fair-weather part of the atmosphere (all those places where no thunderstorms are occurring). It then flows along the earth's surface and back up through the air to the bases of thunderstorms.

Thunderstorms, it turns out, are the whole key to the maintenance of this global circuit. Measurements from aircraft showed the electrical potential in the upper atmosphere to be 200,000 to 300,000 volts with respect to the ground. Combining these values with measurements of conductivity led to estimates that the total worldwide conduction current is something like 1400 amps.

There is no way this global current could continue to flow for more than half an hour or so without being replenished. There had to be an electrical generator somewhere, one that is in almost continual operation.

Thunderstorms, it is now generally agreed, provide the answer. The electrical processes going on inside thunderstorms are complicated, and scientists are divided on exactly what happens. But thunderstorm cumulonimbus clouds do become electrically polarized—positively charged near the top, negatively charged near the bottom.

The sum total of all the thunderstorms in progress on earth is the electrical generator that supplies current to this circuit. The thunderstorms maintain the earth's electric field. If thunderstorm activity were for some reason to stop all over the world, the earth's electric field would fall to near zero in less than an hour. This is an unlikely happenstance. A common estimate is that there are from fifteen hundred to three thousand thunderstorms in progress on the planet at any given moment.

One of the first clues to this global circuit came more than a half century ago when it was noticed that the daily variation in fair-weather electric field intensity measured at the surface varies not with local time but with universal time. This suggested that whatever was happening was not due to local causes but to some common mechanism acting to affect nearly simultaneously the electric field and current all over the globe. The idea that the thunderstorm is the source of this global current gained support when it was noticed that the global electric current variations occurred in step with the times when afternoon and evening thunderstorms would be happening over the continents. There was a readily identified pattern to the variations that seems due to the irregular distribution of the continents rotating with the earth into and out of the afternoon and evening thunderstorm-susceptible time.

Then in 1950 scientists used aircraft to fly directly over the tops of thunderstorms and survey the electric current. They found a positive current averaging a little more than half an amp flowing upward in the clear air above the storms. This has been confirmed many times subsequently. Most scientists now do acknowledge that thunderstorms are the major source of the global electric current.

How is all this related to the sun? There appear to be two effects of the variable sun on the electric field. One, very short term, is a consequence of solar flares. Scientists Reinhold Reiter and William Cobb have shown that within one day after solar flares there is a rapid increase in the electrical current in the global circuit. Ralph Markson, an atmospheric physicist who is a research associate at the Massachusetts Institute of Technology, has speculated that energetic charged particles from the solar flares apparently directly ionize the atmosphere. This reduces the electrical resistance in the air above thunderstorms, allowing greater current to flow upward into the shell of the ionosphere. This results in an increased atmospheric electrical field.

The second effect is longer-term. It involves the obstruction effect on galactic cosmic rays by the solar wind. Remember that the upper atmosphere is conductive primarily due to ionization by this galactic cosmic radiation—high-energy particles from deep space traveling near the speed of light. At times of especially strong solar wind, the abundance of cosmic-ray particles reaching earth is diminished and the atmosphere becomes slightly less conductive.

Markson has studied this matter carefully. He analyzed solar wind data obtained from a series of satellites. He carried out an extensive measurement program of the electric potential in the upper atmosphere using aircraft. From 120 such soundings, he estimated the variation of the ionosphere's electric potential. About half of these measurements were made within an hour of a time when the satellites provided measurements of solar wind velocity.

The result of these comparisons? Clear confirmation of the solar wind effect in reducing the electrification of the atmosphere. The electric potential in the ionosphere had a variation of about one third above and below the average. When solar wind velocity was high, the voltage in the ionosphere was low. When solar wind velocity was low, the voltage in the ionosphere was high. These changes could be seen from one day to the next. The same inverse correlation was seen between values of solar wind and the flow of cosmic rays reaching earth. The high-velocity solar wind apparently does reduce the current in the atmosphere by lessening the amount of ionization from cosmic rays.

Similar long-term measurements using more than 300 soundings from balloons show the same effect. Carried out by University of Minnesota scientist D. E. Olson from 1966 to 1977, they showed that the maximum values of atmospheric current occur at the minimum of the sunspot cycle and the minimum current occurs at the sunspot maximum. Again, that is just what you would expect if solar activity is preventing a portion of the flow of galactic cosmic rays from reaching earth and ionizing the atmosphere.

Markson has long been an ardent proponent of the view that these sun-modulated atmospheric electrification processes are the long-sought physical key to the sun-weather connection.

One beauty of an electrical mechanism is that it could possibly explain sun-weather effects on the time scale of one day. Almost all other suggested associations, such as the variable solar constant, aren't really candidates for changes taking less than about a month. This is partly because the large heat capacity of the oceans helps smooth out and slow down any external heating effect. There are many other such "dampening" effects as well. Changes in solar heating just can't act fast enough to affect weather on very short time scales.

An electrical mechanism also requires little new input of energy. And it neatly skirts the problem of coupling any sun-related effects on the earth's outer atmosphere to the lower atmosphere where the weather takes place. The thin outer part of the atmosphere above 120 kilometers (70 miles) does undergo significant solar-controlled temperature changes, but it is essentially cut off from the weather-producing lower atmosphere. The atmospheric electrical mechanism sidesteps that whole problem. Cosmic radiation affects ionization all the way down to the earth's surface, and solar flare particles can reach deep enough into the atmosphere to ionize the air above the tops of thunderstorms.

What exactly is the link of this atmospheric electrical effect to our weather? To begin with, Markson proposes that the amount of electrical resistance above the world's thunderstorms may in combination act as a kind of valve, regulating the voltage between the ionosphere and the earth and the intensity of earth's electric field. When cosmic radiation or solar flare particles cause greater ionization above thunderstorms, that gate of electrical resistance opens and more charging current can flow. The result is an amplification of the electric-field intensity throughout earth's entire atmosphere, as Markson puts it, "from ionosphere to the ground and from pole to pole."

All this so far is on fairly firm footing. The next step, a suggested sun-caused link to actual weather processes, is more speculative. Markson suggests that this amplified fair-weather electric field may in turn stimulate more electrification in clouds. This would be a positive-feedback effect. The electrical activity in thunderstorms charges up earth's electric field (in collaboration with the sun and cosmic rays), and this intensified electric field in turn causes even more electrical activity in existing thunderstorms and initial electrification of newly developing cumulus clouds, which may then grow into more full-fledged thunderstorms.

It is quite a grand conception, this global circuit and this proposed feedback amplification cycle.

The effect of electric fields on the physical processes in clouds is not well understood. Markson notes that experiments and computer modeling indicate that electric fields may help stimulate the electrification of thunderstorms and they may also play a role in the condensation of water vapor into droplets and in the coalescence of droplets into raindrops. If so, they would affect both rainfall and the addition of heat to the atmosphere. The heat can cause upward convection that can generate more storminess. All sorts of complex effects can take place. Scientists disagree on the importance of atmospheric electricity on cloud physics, but Markson finds it difficult to imagine that intensified electric fields do not have some effects on the physical processes going on inside storm clouds. Any effect on thunderstorm activity would be important, for, as Markson notes, with some understatement, "thunderstorms are of major meteorological importance over much of the earth's surface."

After a succession of published reports on this work in 1978, 1979, and 1980, Markson completed significant additional work. He directly correlated cosmic radiation with the only available data sets of ionospheric potential variation and found the expected direct correlation. So now, in his view, there seems to be little doubt that the proposed mechanism is correct, although he and colleagues continue to work with statistical correlations and want more explicit proof. The implication of the direct correlation is that the thunderstorm charging mechanisms are sensitive to conductivity over developing and mature thunderclouds and the electric field intensity near them. A second implication is that there probably is indeed an increase in thunderstorm activity following increases in ionization of the atmosphere.

Markson's work offers an explanation for how solar variability controls the electrification of earth's atmosphere. And it proposes ways that the changing sun may in turn affect weather. The long-sought physical link may be at hand. "It is no longer necessary to investigate the sun-weather problem only through statistics and correlations," he says. "We now have the tools to observe the details of chains of events by which solar activity may influence meteorology in the lower atmosphere."

Sun-weather scientists look on the atmospheric electricity mechanism with considerable interest. That interest is mixed with the caution born of a realistic view of the difficult history of this quest for a sun-weather connection.

Walter Orr Roberts proposed a different theory involving the action of charged particles on the aurora. The particles would change electrical conditions in the atmosphere and encourage the formation of large heat-trapping cirrus clouds. Unfortunately, this mechanism can't act fast enough to account for the perceived quick (twelve-hour) response of weather patterns to solar sector boundary crossings. He now considers that idea dead.

Markson's idea, he says, is the "only other hypothesis that seems to me to make sense." As Roberts told me, "Markson's proposal is very suggestive. It is most likely to be right of all those on the table."

Everyone, Roberts included, agrees that the idea needs more testing. One way is with computer models. Scientists at the National Center for Atmospheric Research in Boulder have developed a numerical model to study the global circuit. Such studies are a method of mathematically testing a theory, a computer simulation of nature.

The model has shown that changes in the electrical conductivity of the upper atmosphere resulting from such events as solar flares do in fact alter the global distribution of electrical fields and currents. It suggests that solar activity does indeed produce a small but measurable effect on the global electrical current. This of course bolsters the electrical link as a candidate for the sun-weather mechanism.

Studies reported in late 1979 by NCAR atmospheric physicist J. Doyne Sartor, who died only a short time later, show that these electrical-field intensifications do have remarkably strong effects on the growth and circulation of some storms. His work demonstrated that under certain conditions, sun-modulated disturbances of the electric field should have considerable effect on storm activity. These electrical effects, his computer studies showed, may even be as strong as the atmospheric dynamic processes usually invoked to explain the generation of storms.

He concluded that when the electrical current in the atmosphere becomes more intense due to solar flare ionization, the electrical charge in clouds is intensified. "These stronger fields and charges," he reported, "speed up precipitation growth processes and produce vertical and horizontal vorticity in thunderstorms and their anvils and in some highly electrified mesoscale clouds (such as tropical clouds) comparable to the production of vorticity by atmospheric dynamics." It was just as Markson had suggested. Sartor concluded that it was possible that the intensification of earth's electric field caused by solar flares might make the difference between a weakly and a strongly electrified storm.

These computer studies lend strong credibility to the atmospheric-electrical sun-weather mechanism. What is really needed, however, is continual measurement of the electrical state of earth's atmosphere. We do not have it.

Markson suggests that just as we routinely monitor electrical conditions on the sun and in space, "it would seem reasonable to continue such measurements within the earth's atmosphere." Such measurements would establish a "geoelectric index" comparable to the geomagnetic indexes that have been indispensable to solar-terrestrial studies during this century.

Adding instruments to measure the vertical electric field, conductivity, and the production rate of ions to a few of the radiosonde meteorological balloons launched by the National Weather Service twice each day would certainly help. Even that is not now done.

Far preferable would be continuous measurements of the variations in the atmosphere's electrical state. Markson suggests that aircraft or perhaps balloons or kites tethered to ships at sea might be the best way to monitor variations of electric potential in the atmosphere. The instrumented balloons or kites would have to fly at altitudes of at least 5 kilometers (16,000 feet) to do the measurements. Measurements of the atmosphere's electrical conductivity and ion-production rate would have to be made from free-flying balloons 25 to 30 kilometers up in the stratosphere.

All this involves existing technology. Such measurements could clarify exactly how solar variability causes atmospheric electric effects and go a long way toward showing how they in turn do or do not affect the weather. An atmospheric electrical effect on the electrification of clouds, the intensification of rainfall, and the dynamic growth processes of clouds seems the most likely way for solar activity to rapidly stimulate changes in the atmosphere and the weather. It would at least be wise to test the ideas as fully as possible.

The problem is that any continuing measurement program, even when simple in concept, is expensive. I asked a National Science Foundation scientist involved in that agency's solar-terrestrial research program about Markson's suggested measurement program. He said Markson's entire concept is interesting, but the program of measurements he has proposed "tends to be controversial," mainly although not entirely due to its expense. Markson has proposed a program to measure ionospheric electrical potential daily for one year at a cost of half a million dollars. "That's a lot," the NSF official said. "We like to see proposals in the one-hundred-thousand-dollar area." Despite certain statements by politician-critics, money for scientific projects seldom comes easily. There is always so much that is worthwhile to do. Nevertheless scientists usually are fairly resourceful in managing somehow to get at least some of the measurements they need, and we might hope that will eventually prove true in this case.

As Markson said in his 1980 report in *Science*: "The question of how solar variability affects the atmosphere is most timely as we turn to the sun and space for possible answers to the problems resulting from increasing population in a world with limited food, energy, and natural resources—all of which are directly or indirectly affected by atmospheric conditions. Thus if solar activity affects meteorological processes, and there is increasing evidence suggesting that it does, and if society is going to develop the capability of predicting and rationally planning its future, the sun-weather problem becomes a matter of considerable importance."

13. THE DECADE OF THE SUN

The Cup dipped into the sun. It scooped up a bit of the flesh of God, the blood of the universe, the blazing thought, the blinding philosophy that set out and mothered a galaxy, that idled and swept planets in their fields and summoned or laid to rest lives and livelihood.
— Ray Bradbury, *The Golden Apples of the Sun*

Perhaps sometime in our spacefaring future we will actually visit the sun, scooping up some star matter as Bradbury envisioned for the crew of the sun-skimming spaceship *Copa de Oro,* also named *Icarus.* But we need not go to the sun to sample the sun. The stuff of the sun, we now know, is flung out across the solar system. It comes to us; we need not go to the source.

Yet there is a kind of observation of the sun that we earthlings, despite nearly four centuries of telescopic peering and a quarter century of spacecraft monitoring, have never achieved. Space explorers we may be, with our robot, computer-controlled remote eyes making wonderful close-up discoveries about the many astonishingly beautiful and diverse objects in the solar system. We have scanned the crater-wracked surface of Mercury. We have dropped probes down through the swirling thick atmosphere of our blazing hot neighbor planet Venus. With radar on our Pioneer satellite in Venus orbit we have mapped its cloud-shrouded surface. We have landed on wind-swept Mars, probed its surface for life, studied its strange soil chemistry, and charted its vast canyons channeled out by ancient flows of water now locked up for epochs of time in the Martian polar ice cap. We have discovered whole new worlds in our missions past the four giant moons of Jupiter, each distinct from the others, one splashed with color like a Jackson Pollock painting, another venting forth plumes from volcanic eruptions in progress. We have photographed up close the colorful bands and swirling red spot on Jupiter, a giant planet itself a once would-be star that failed to ignite. We have seen lightning superbolts and auroras on Jupiter, and have even recently recorded the flash of a fireball as a meteor met death plunging into the top of its atmosphere. We have flown directly beneath the rings of Saturn and shown them to consist of hundreds of finely etched ringlets like some exquisite art deco drawing. And our own moon-bound astronauts have looked back at home planet earth and seen a beautiful, fragile gem set in the lonely darkness of space. We have found the solar system an aesthetically wondrous place worthy of the exploratory quest that seems built deeply into the nature of our species.

In all these explorations, however, we have been confined more or less to the plane of the solar system, much like Mr. Tompkins in Flatland. Our spacecraft eyes have been a few degrees above and below the plane, but not much more than that. Our view of sun, in particular, has been almost entirely two-dimensional. We have had a hint of the extraordinary importance of the three-dimensional structure of the sun's output of magnetic fields and particles. That has come from the newfound conception of the heliospheric current sheet I described in chapter 11, radiating outward from the sun and warped in three dimensions like the petals of a flower rippling in the wind.

Our view of this elegant if invisible structure has come from observations made only within 16° of latitude from the sun's equator, a pinnacle obtained by our Venus-bound Pioneer 11 spacecraft. We've only directly observed a fairly thin slice of it, in other words. All evidence indicates that at solar maximum the warped current sheet reaches latitudes 50° or more above and below the plane of the sun's equator. We know also that the sun's coronal holes (chapter 2) exist in their most permanent form at the sun's polar regions. We detect only a portion of the solar wind emanating from polar holes, most generally when the holes expand equatorward during the declining phase of the solar cycle. The solar wind that flows out high above and below the sun-earth plane goes unnoticed.

How special to be the first person to see the sun and our solar system from above! A perspective never before possible in human history! This decade should bring us that view. No human will be there in person, but an unmanned spacecraft will ably substitute for us. It is called the Solar Polar mission. It was to be a truly international effort, with two spacecraft involved, one provided by the United States (NASA), the other by the European Space Agency. But in March 1981 the budget-conscious Reagan administration cut off funding for the U.S. spacecraft, much to the displeasure of the Europeans and the American space community. It appeared as though the Europeans would have to go it alone, although there would be U.S. instruments aboard their craft, NASA would help provide navigation and support services, and the craft would be launched into space by the Space Shuttle, and then be sent off to the sun by way of Jupiter.

Why Jupiter? That's a little like going from Cheyenne to Denver via New York City. Well, Jupiter's enormous mass will provide the gravitational energy to swing the spacecraft up out of the plane of the solar system. Without that kind of boost, our space vehicles don't have enough power to escape out of that plane. Jupiter will not only accelerate the two solar-polar craft out of the grip of that plane but send them on a trajectory toward the sun.

In the two-craft plan, one was to fly past the sun's north pole and then the south pole. The other would do just the reverse. The instruments on board the two craft were complementary. In the revised plan, only the one craft will be available to attempt to chart the three-dimensional structure of its corona and the three-dimensional distribution of the charged particles, magnetic fields, and high-energy electromagnetic emanations (such as X rays and gamma rays) the sun spews out across interplanetary space.

If no further problems develop, the European spacecraft will be launched in 1985 and spend about a year above ecliptic latitude 40° in 1988 and 1989. The first-ever out-of-the-ecliptic views of the sun they will provide should be exciting.

Another mission, now only on the drawing boards for a future possible launch in the late 1980s, would come closest to fulfilling the fictional feats of Ray Bradbury's brave crew of the *Copa de Oro*. It would be our closest journey to the sun. Not that the spacecraft would have a crew of human beings—there is really no need for that. The Solar Probe, as this proposed spacecraft is called, would be another automated, unmanned robot. The scientists in our Viking and Voyager projects have shown how sophisticated such probes can be.

Think of the view it would transmit back to earth! In the mid-1970s two German sun-monitoring craft, Helios 1 and 2, reached to about 70 percent of the distance from earth to the sun. Solar Probe would close to within 98 percent of the earth-sun distance, whipping around the sun barely 2 to 3 million kilometers above the sun's surface. The Solar Probe mission has been aptly called "an encounter with a star."

The heat would be the intensity of 2,500 suns shining on earth. A special heat shield would be kept pointed at the fiery surface to protect the instruments. Solar Probe's imaging systems in visible, ultraviolet, and X-ray light would provide resolutions 5 to 50 times better than earth-orbiting telescopes. Other instruments would measure magnetic fields and chart the source of the solar wind with a precision never before possible. And precise tracking of the probe's motion as it passes by the sun would allow scientists to carry out tests of how the sun's mass is distributed in its interior. They should be able to tell whether part of the sun's interior is slightly flattened rather than spherical, an effect expected if the core of the sun is rotating faster than the rest of the sun. The dynamics of the sun will be better understood.

The warping of space around the sun due to its enormous mass (as explained by Einstein's general theory of relativity) should produce measurable effects on the probe's motion that will provide a further experimental test of relativity. Solar Probe will be our civilization's closest brush with the sun. The quest that began with Icarus will have been fulfilled.

The decade of the 1980s began with an ambitious space-based program to monitor the sun. Solar cycle 21 peaked during late 1979 and 1980. The Solar Maximum Mission was timed to coincide with that high point of activity. The SMM satellite was launched February 14, 1980, and it was like a valentine to solar scientists. For one thing, as I reported in chapter 7, within five months it had revealed that the sun does exhibit variations of a few tenths of a percent in its output of light over periods as short as a week. The long-sought quest to show that the sun is a variable star was being put on a precise quantitative basis.

Solar Max's primary goal, however, was to observe solar flares from birth to death, and that it did with spectacular success. The satellite kept the sun in view for 60 out of every 96 minutes it takes to orbit earth. Early each afternoon the scientists responsible for its work met in a room at the Goddard Space Flight Center in Greenbelt, Maryland, to plan observations for the twenty-four-hour period beginning the following morning. They had the benefit of daily solar weather forecasts from the government's Solar Forecast Center in Boulder plus instantaneous views of the sun transmitted from solar observatories such as Big Bear in California and Sacramento Peak in New Mexico. While some of the satellite's instruments kept the whole sun in view, others focused in on particular regions that seemed likely to erupt into a flare.

One giant flare caught in action for 40 minutes on May 21, 1980, resulted in what the scientists at the time called "the most significant observations ever made of a solar flare." The flare disrupted earth's ionosphere sufficiently to cut off terrestrial radio communications over a broad band of frequencies for more than a half hour. It stimulated a nine-hour geomagnetic storm on earth. Fortunately there were few other effects.

The May 21 event offered the first opportunity for Solar Max's telescopes to focus on the hottest part of a giant flare. The readings seemed to indicate enormous temperatures at the hottest moment of a flare, more than 56 million degrees Celsius, hotter than the inside of the sun and far hotter than the normal 4,400° to 5,500° C on the solar surface. These temperatures are higher than anyone had thought was possible on the sun. Even more surprising is that the gaseous matter in them didn't expand. The density increased, not decreased, with the sudden rise in temperature, showing that powerful forces, presumably magnetic, keep the high-temperature gas well confined. That's something that applied physicists carrying out research into future thermonuclear fusion reactors for energy on earth are striving to learn how to do.

One flare under observation even before its birth confirmed one theoretical idea about how flares are generated. The flare occurred when a rising loop of ionized gas and magnetic fields collided with a higher lying loop. The collision seemed to release energy seen as a flare. This collision mechanism would explain why flares aren't produced in every magnetic region on the sun. Perhaps this kind of interaction is required.

During the period May 15 to June 30, 1980, scientists all over the world cooperated in gathering as many scientific measurements of the sun from as many sources as possible. These dates had been established two years earlier, when scientists gambled that the sun would be prominently active during that particular six-week period. They weren't disappointed.

On June 24, as a result of predictions from the worldwide solar forecast services, all of the flare instruments on board the SMM spacecraft, as well as instruments at observatories in eighteen countries, were focused on solar active region 2522. That day and the next, two major flares erupted from region 2522.

All seven of Solar Max's flare instruments caught the flares in action, the first time that had happened. The flares occurred when the sun was in view of the United States, South America, and Australia. Observatories in Brazil, Australia, New Mexico, California, and Hawaii all had them under observation.

At the new Very Large Array radio telescope west of Soccoro, New Mexico, for example, the precise location where electrons accelerated in the flare emit radio waves was detected. When combined with the location information on the same particles gathered by the SMM satellite's X-ray and gamma-ray instruments, the scientists obtained the most detailed observations ever made on the release of energy in a flare during its earliest stages.

The two flares turned out to be what scientists call homologous. They were nearly identical in structure, position, and character. This could be sheer coincidence, but another possibility is that the magnetic field returns to exactly the same configuration after one flare erupts. The instability that triggers the second flare is thus almost identical to the first.

All this is the kind of new information scientists love. It is the raw material out of which new advances in understanding nature come. The Solar Maximum Mission satellite's work in 1980 got the decade—I call it the Decade of the Sun—off to a rousing start.

Why the Decade of the Sun? It is not only because of the spacecraft missions I've been describing. Spacecraft (including the Space Shuttle, which will carry the Space Telescope, the Spacelab, and other sun-monitoring equipment into orbit) are an important part of it. The new perception of a changing, irregular sun has stimulated an enormous amount of new scientific and public interest in the sun.

That heightened interest coincides with a host of new ways of studying the sun, just now arriving on the scene. Together, they provide the makings of a new revolution in solar science.

Solar scientists believe the time is now ripe for a thrust forward. They feel the 1980s will, for example, bring significant advances in the study of the fundamental mechanisms responsible for the solar cycle. And if we understand that, we'll better understand what's responsible for the kinds of irregularities in the cycle that both intrigue and trouble us.

The 1980s should provide better answers to many questions about the solar cycle. How exactly does the solar dynamo operate to produce a single twenty-two-year magnetic cycle? How does the sun's convection zone, where magnetic fields and solar activity originate, change during a cycle? How do these changes affect the sun's output of radiation, charged particles, and magnetic fields? What is happening in the convection zone and with the solar dynamo to account for the long-term excursions in solar activity such as the Maunder minimum? And beyond the short-term solar cycle, what is the long-term behavior of the sun all the way back to its ancient past, and what can we infer from all that about its behavior in the near future?

You have to admit that is a fairly significant set of questions. Yet there is a very real possibility of answering some of them in this decade. Why? One of the important ways of studying a dynamic system like the sun is what scientists call computer modeling. You plug all of the relevant known information about the sun into a "model"— a set of equations representing the physical principles we believe should be governing events—and figuratively set it all in motion. Then you watch what happens, again via the numerical output of the computer. With the large computer, we now have the chance to develop realistic numerical models for the motions of the hot, charged gases in the sun's all-important convection zone. From them we can calculate the effects of these motions on the solar dynamo. Such models will also make certain predictions about the operation of the solar cycle that we can then compare to the behavior of the sun itself and thus advance our understanding of the true nature of the solar cycle. As these computer models become more sophisticated, they should be able to reproduce more and more of the well-known characteristics of the sun's cycle. Then perhaps we can extend our knowledge into past eras. Eventually, we might be able to ask the computer how the sun's output of radiation differed during such periods as the Maunder minimum.

Solar scientists are also excited by the advent of solar seismology. They are on the verge of gaining unprecedented information about the inside of the sun. This should be the reward of observing the different oscillations coursing through the sun in such fine detail that vibrations with only slightly different time periods can be distinguished from each other (chapter 5). Observations from space made possible by the Space Shuttle in the 1980s should provide the needed resolution. Observations from such places as the South Pole, where the summer sun can be kept in continuous view for many days and nights, will also continue to contribute significantly to that quest. The solar scientist is just about to get the tools for seeing into the sun that the earth scientist has for studying the internal structure of our planet by observing seismic waves from earthquakes.

Ultrasensitive velocity-measuring devices can now be built that will be able to detect giant vertically moving cells of electrified gas on the sun's surface. Precise temperature-measuring instruments will be able to detect the presence of such rising and descending eddies by their slight differences in temperature. These long-lived, large-scale circulation cells are like the ascending columns in a pot of boiling water. They and the magnetic patterns associated with them should carry an indelible signature of deep-seated circulation. That circulation in turn is tied intimately to the fundamental operation of the solar cycle.

We are also getting close to replacing the question "Does the sun's output of radiation—the solar constant—vary?" with *"How much* and *when* does the solar constant vary?" The instruments on board the Solar Maximum Mission spacecraft which showed in 1980 that the solar constant changes a few tenths of a percent in periods as short as a week (chapter 7) will be further improved throughout the decade. They should reveal not only the influence of individual sunspots but also the effects of the solar cycle itself. Thus, as solar physicist Gordon Newkirk points out, "The long-sought answers to the questions of whether climatologically important changes in the solar flux occur and what their solar origin might be may well be at hand."

We'll also be looking into the distant reaches of space for clues about the likely behavior of our own sun. It is exciting to be able to tell that other stars with masses similar to the sun have activity cycles like those we see on the sun. We can infer the eruption of massive flares on distant suns and the presence of abundant starspots, just like our star's sunspots, on other stars. This is all very new work, and the 1980s will see many exciting developments in this kind of comparative solar science. We will vastly improve our view of our changing sun in the cosmic perspective.

We will also see dramatic gains in our record of the sun's past behavior. As lunar scientist Bevan French has pointed out, the record of the sun's past is found almost everywhere in the solar system. Scientists from a variety of backgrounds are beginning to accelerate the search for these recordings of solar variability. Some of them go all the way back to the sun's ancient beginnings. The moon, for instance, is a virtual museum of past solar activity, its rocks containing the etchings of eons of bombardment by energetic particles from solar flares and the solar wind. Its riddled and churned-up surface has been shaped by the sun. As space scientist Robert M. Walker said at a recent Conference on the Ancient Sun, "When you pick up a handful of lunar dust you are picking up stardust—the star being the sun."

Some of these solar etchings were made during the very earliest periods of the solar system's history and incorporated inside of conglomerate lunar rocks, where they are protected from erosion by micrometeorites striking the moon. Meteorites, fragments of rock that orbited the sun from very early times before happening to collide with the earth, also contain such records. Deciphering the record correctly is difficult, but the very coarse picture so far is comforting. It shows, for instance, that over the very long view the behavior of the sun apparently was not so much different in the ancient past from what we see now. Particle tracks of solar flares have been found in meteorites that may be as old as 4.2 billion years. And the inferred energy of these flare particles seems not to have changed too much during all that time. So the sun seems to have been perking along over the eons much as it does now.

We shouldn't take too much comfort from this kind of evidence, however. It has large uncertainties, and the scientists working with it admit that fact. Other kinds of studies do show evidence of significant variations. Solar flare particles can produce certain radio-active isotopes in lunar rocks, and some of the rocks brought back by the Apollo astronauts reveal this record. One recent study of the abundance of an isotope called krypton-81 in a rock found by the Apollo 12 crew implies that for a time several hundred thousand years ago the sun was generating solar flare protons in quantities two or three times greater than either before or since. There is also some similar kinds of evidence for higher solar flare activity ten thousand or so years ago.

These kinds of discoveries, although they are far from being confirmed, are troubling. What the sun has done in the past it might also do in the future, and such a bolstered output of solar particles could conceivably have effects on climate and even life.

Another very real mystery about the sun has just recently surfaced from our studies of the moon's soil. The chemical composition of the solar wind has apparently changed over the history of the moon (and sun). The ratio of two forms of nitrogen has increased by at least 30 percent. No one quite knows what to make of this discovery. It is disturbing and intriguing. Apparently, for whatever reason, nuclear reactions on the surface of the sun (not the energy-releasing reactions in the sun's thermonuclear core) produced in the past a heavier form of nitrogen in quantities that the sun is not now capable of generating. Something important seems to have changed on the sun. It's all a puzzle.

This study of the history of the changing sun by way of moon rocks, lunar soil, and meteorites is sure to produce more surprises in the coming years.

Yet we've hardly exhausted the record of solar variability right here on earth. The sun-modulated carbon-14 content of tree rings has been a boon to study of past solar behavior, but even more kinds of chemical tracings of changes in the sun are frozen into our planet. I mean "frozen" quite literally. Take the nitrates found in the ice of Antarctica, for example.

Nitrates are produced in the upper atmosphere over the polar regions when solar particles interact with atoms in the air and generate auroras. These nitrates become incorporated into the annual snowfall over the Antarctic ice cap, where, due to the absence of any summer melting, each year's layer of snow is preserved. Once all other sources of nitrates are ruled out, the nitrate composition in the ice then becomes a signature of past solar activity. When solar activity is high (and auroras abundant), more nitrate is produced. When solar activity is low (and auroras infrequent), less nitrate is produced. Edward Zeller of the University of Kansas and Bruce Parker of Virginia Polytechnic Institute say this nitrate record of solar activity can be read just like tree rings. Each year's layer of ice is distinguishable. Even better, the nitrate fallout becomes incorporated into the ice quite quickly (within one or two years), whereas carbon-14 becomes absorbed into the wood of trees only after a time lag of thirty to fifty years. So the nitrate record is theoretically more precise.

Using these techniques on ice cores from the South Pole and from the Russian station at Vostok, 1,500 kilometers (930 miles) from the Pole, the two scientists believe they have indeed begun to find clear signatures of past solar activity. The nitrate record closely follows, with the expected two-year time lag, the known record of recent sunspot activity. With the confidence gained from that connection, the scientists go further back in time and find a clear trace of—you guessed it—the Maunder minimum of solar activity. During that time (A.D. 1645 to 1710) nitrate concentrates drop off from an average of about 23 parts per billion to about 10 and sometimes 6 parts per billion.

There is every reason to believe that the nitrate record of solar activity is preserved in ice cores back many tens of thousands of years. This work, now just in its embryonic stages, is sure to produce important findings about the sun over the next decade.

Scientists also place hope in another chemical tracer of past solar activity. This one is a radioactive isotope, called beryllium-10. Like carbon-14, it is produced in the earth's upper atmosphere by the action of galactic cosmic rays. When solar activity is high some of the cosmic rays are kept from impinging on earth, and fewer atoms of both carbon-14 and beryllium-10 are formed.

All this has been known for some time. The problem with beryllium-10 is that it decays only very slowly—its half-life is 1.5 million years—and that makes it very difficult to detect. That problem has been solved by the development of new techniques using particle accelerators that detect not the products of radioactive decay but the actual atoms of beryllium-10 itself.

Beryllium-10, unlike carbon-14, precipitates out of the atmosphere quite rapidly and becomes incorporated into handy repositories such as the ice caps of Antarctica and Greenland and the sediments at the bottom of the oceans.

A French scientist, Grant Raisbeck, has been developing this proxy indicator of solar activity. Even though he has had trouble obtaining ice cores for analysis, he recently did complete preliminary studies on a core taken by French scientists from a site known as Dome C in East Antarctica. The first thing Raisbeck did was look for evidence of the Maunder minimum. Although he is cautious about stating his conclusions, his graph shows clear evidence for it. Beryllium-10 concentrations are notably elevated during that time interval.

The long lifetime of beryllium-10 makes possible its use as a proxy indicator of the sun's past behavior back several tens of millions of years. An enormously thick history book of the sun may be about ready to open before our very eyes. Solar scientists are understandably enthusiastic.

Still another possible indicator of past solar cycles may be preserved in annual depositions of glacial debris on ancient lake bottoms. An Australian geologist, R. E. Williams, reported in June 1981 that he has found cycles of 11 years, 22 years, 90 years, and even longer periods in the thickness variations of such varve layers in a formation of ancient siltstone and sandstone in southern Australia. This deposit is of Precambrian age, more than 600 million years old, long before the dawn of the ages of fishes, amphibians, and reptiles. If Williams is right in supposing that these layers represent a sun-caused cycle of climate variation, this is the most ancient known evidence of the solar cycle that I have ever heard of. It would suggest that for some reason, perhaps a weakening of the earth's magnetic field, solar control of climate was much greater in Precambrian times than it is today. And, as Williams said, such "ancient annually layered rocks should be seen as potential solar observatories that may extend our understanding of solar processes and solar-planetary relations."

I find it satisfying to think that there are so many intriguing ways the earth manages to capture and preserve a record of past solar variation. I think our learning how to read some of these clues says less about human ingenuity than it vividly emphasizes the astonishingly intricate relationships there are in nature among things that on the surface seem to have no connection. The sun, cosmic particles from elsewhere in the universe, atoms in earth's upper atmosphere, the chemicals in ice cores from the frozen wastes of Antarctica, and (with carbon-14 in tree rings) the very intimate processes of life itself—

all are intertwined, creating a kind of time capsule that our human eyes and minds are only now beginning to see and appreciate.

The Zuñi Indians have a prayer offering to the rising sun:

Now this day,
My sun father,
Now that you have come out standing to your
*   sacred place,*
That from which we draw the water of life,
Prayer meal,
Here I give to you.

The sun is life. Traditional cultures everywhere are quite right to revere it. They well know that all life would quickly end without the bountiful munificence of its rays. They feel a connection with the sun that most of us, in the confines of our homes and cities, have lost. The most fundamental alteration of the sun that affects human civilization has always been its seasonal change in position in the sky. That has nothing to do with the sun itself, but only with the progression of our tilted planet along its annual orbit about the sun. Our modern society is privileged to have the means of understanding the sun as not a featureless ball of fire in the sky but as a star, one so close that it not only nourishes all life on earth but serves as a laboratory for understanding the star-strewn universe. It gives us perspective on the cosmos.

The past half-dozen years have seen the rise of profound new questions and mysteries about this home star of ours. With our new ways of learning about the sun we have become aware of previously unsuspected subtleties and never-before-detectable variations. Our perception of the sun has been sharpened. Our appreciation of the sun has been regenerated. We now know we live in the streams of a changing star. What the future of this quest will reveal no one can tell. One thing is certain: the sun's store of enigmas has not been exhausted.

SELECTED BIBLIOGRAPHY

I have added slight annotations to the listings of several works of special interest.

Chapter 1. A SUN FULL OF SURPRISES

Bray, R. J., and R. E. Loughhead. *Sunspots.* London: Chapman and Hall, 1964.

Eddy, John A., ed. *The New Solar Physics.* Boulder, Colo.: Westview Press, 1978.

Frazier, Kendrick. "The Sun Our Star." *Science News,* 113:252–54 (1978).

Friedman, Herbert. *The Amazing Universe.* Washington, D.C.: National Geographic Society, 1975. Fine section on the sun.

Meadows, A. J. *Early Solar Physics.* New York: Pergamon Press, 1970. Brief history of nineteenth-century solar astronomy with reprints of many seminal papers.

Menzel, Donald H. *Our Sun.* Cambridge, Mass.: Harvard University Press, 1959. Readable text on classical sun.

Moore, Patrick. *The Sun.* New York: Norton, 1968.

Pasachoff, Jay M. *Contemporary Astronomy,* 2nd ed. Philadelphia: W. B. Saunders Co., 1981. Well-written text with extensive section on the sun.

Pasachoff, Jay M. "Our Sun." *Astronomy,* January 1978, pp. 6–24.

Parker, E. N. "Solar Physics in Broad Perspective." In *The New Solar Physics,* ed. John A. Eddy. Boulder, Colo.: Westview Press, 1978.

Parker, E. N. "The Sun." *Scientific American,* September 1975.

Zeilik, Michael. *Astronomy: The Evolving Universe,* 3rd ed. New York: Harper & Row, 1982. Excellent, up-to-date text.

Chapter 2. THE SUN'S IMPACT ON EARTH

Akasofu, S.-I. *Aurora Borealis: The Amazing Northern Lights.* Anchorage: Alaska Geographic Society, 1979. Beautifully illustrated, up-to-date treatment by noted authority.

Akasofu, S.-I. "Solar Cycle Review." In *Physics of Solar Planetary Environments,* ed. Donald J. Williams. Washington, D.C.: American Geophysical Union, 1976.

Bohlin, J. D. "The Physical Properties of Coronal Holes." In *Physics of Solar Planetary Environments,* ed. Donald J. Williams. Washington, D.C.: American Geophysical Union, 1976.

Carrington, R. C. "Description of a Singular Appearance Seen in the Sun on September 1, 1859." *Monthly Notices of the Royal Astronomical Society* 20:13–15 (1860).

Eddy, John A. *A New Sun: The Solar Results from Skylab.* NASA SP-402. Washington, D.C.: U.S. Government Printing Office, 1979. Lavishly illustrated book for the general reader.

Krieger, A. S., A. F. Timothy, and E. C. Roelof. "A Coronal Hole and Its Identification as the Source of a High Velocity Solar Wind Stream." *Solar Physics* 29:505–25 (1973).

Maran, Stephen P. "Counting the Sunspots." *Natural History,* August–September 1979, pp. 111–16.

McKinnon, J. A., et al. "August 1972 Solar Activity and Related Geophysical Effects." NOAA Technical Memorandum ERL SEL-22, National Oceanic and Atmospheric Administration, Space Environment Laboratory, Boulder, Colo., 1972.

Sargent, H. H., III. "Very Large Geomagnetic Disturbances During Sunspot Cycle 21: A Prediction." In *Solar-Terrestrial Predictions,* Vol. 2, ed. R. F. Donnelly, National Oceanic and Atmospheric Administration, Space Environment Laboratory, Boulder, Colo., 1979.

Smith, R. Jeffrey. "The Skylab Is Falling and Sunspots Are Behind It All." *Science* 200:28–33 (1978).

Wilcox, John M. "History of Solar-Terrestrial Relations as Deduced from Spacecraft and Geomagnetic Data: Solar M Regions." In *Physics of Solar Planetary Environments,* ed. Donald J. Williams. Washington, D.C.: American Geophysical Union, 1976.

Chapter 3. THE INCONSTANT SUN

Eddy, John A. "The Maunder Minimum." *Science* 192:1189–1202 (1976).

Eddy, John A. "The Sun Since the Bronze Age." In *Physics of Solar Planetary Environments,* ed. Donald J. Williams. Washington, D.C.: American Geophysical Union, 1976.

Eddy, John A. "The Case of the Missing Sunspots." *Scientific American,* May 1977, pp. 80–92.

Eddy, John A., Peter A. Gilman, and Dorothy E. Trotter. "Anomalous Solar Rotation in the Early 17th Century." *Science* 198:824–29 (1977).

Feynman, J., and S. M. Silverman. "Auroral Changes During the Eighteenth and Nineteenth Centuries and Their Implications for the Solar Wind and the Long-Term Variation of Sunspot Activity." *Journal of Geophysical Research* 85: 2991–97 (1980).

Herr, Richard B. "Solar Rotation Determined from Thomas Harriot's Sunspot Observations of 1611 to 1613." *Science* 202:1079–81 (1978).

Maunder, E. W. "Professor Spoerer's Researches on Sun-Spots." *Monthly Notices of the Royal Astronomical Society* 50:252–52 (1890).

Siscoe, George L. "Evidence in the Auroral Record for Secular Solar Variability." *Reviews of Geophysics and Space Physics* 18:647–58 (1980).

Stuiver, Minze, and Paul D. Quay. "Changes in Atmospheric Carbon-14 Attributed to a Variable Sun." *Science* 207:11–19 (1980).

Waller, William. "Astronomers Compare Our Sun with Other Stars." Smithsonian Institution *Research Reports* No. 28:5–7 (1980).

Wilson, O. C. "Chromospheric Variations in Main-Sequence Stars." *Astrophysical Journal* 226:379–96 (1978).

Wilson, Olin C., Arthur H. Vaughn, and Dimitri Mihalis. "The Activity Cycles of Stars." *Scientific American,* February 1981, pp. 104–19.

Chapter 4. A PAUCITY OF GHOSTS

Bahcall, J. N. "Solar Neutrino Experiments." *Review of Modern Physics* 50:881 (1978).

Bahcall, John N., and Raymond Davis, Jr. "Solar Neutrinos: A Scientific Puzzle." *Science* 191:264–67 (1976).

Bahcall, John N., et al. "New Solar-Neutrino Flux Calculations and Implications Regarding Neutrino Oscillations." *Physical Review Letters* 45:945–48 (1980).

De Rujula, A., and S. L. Glashow. "Neutrino Weight Watching." *Nature* 286:755–56 (1980).

Gamow, George. *Thirty Years That Shook Physics.* Garden City, N.Y.: Anchor/Doubleday, 1966.

Hartline, Beverly Karplus. "In Search of Solar Neutrinos." *Science* 204:42–44 (1979).

Morrison, Philip. "The Neutrino." *Scientific American* January 1956, pp. 58–68.

Thomsen, Dietrick E. "Weighed in the Balance and Found: Neutrino." *Science News* 117:292–93 (1980).

Thomsen, Dietrick E. "Whatever Happened to Neutrino Mass?" *Nature* 287:481 (1980).

Chapter 5. THE SHAKING SUN

Brookes, J. R., G. R. Isaak, and H. B. van der Raay. "Observation of Free Oscillations of the Sun." *Nature* 259:92–95 (1976).

Brown, Timothy M., Robin T. Stebbins, and Henry A. Hill. "Long-Period Oscillations of the Apparent Solar Diameter." *Astrophysical Journal* 223:324–38 (1978).

Claverie, A., et al. "Solar Structure from Global Studies of the 5-Minute Oscillation." *Nature* 282:591–94 (1979).

Deubner, F. L., R. K. Ulrich, and E. J. Rhodes, Jr. "Solar p-Mode Oscillations as a Tracer of Radial Differential Rotation." *Astronomy and Astrophysics.* 72:177–85 (1979).

Grec, Gérard, Eric Fossat, and Martin Pomerantz. "Solar Oscillations. Full Disk Observations from the Geographic South Pole." *Nature* 288:541–44 (1980).

Hill, Henry A. "Seismic Sounding of the Sun." In *The New Solar Physics* ed. John A. Eddy. Boulder, Colo.: Westview Press, 1978.

Hill, Henry A. "Solar Pulsations." Paper presented at IAU Colloquium No. 58, "Stellar Hydrodynamics," in Los Alamos, N.M., August 12–15, 1980.

Hill, Henry A., and Thomas P. Caudell. "Global Oscillations of the Sun: Observed as Oscillations in the Apparent Solar Limb Darkening Function." *Monthly Notices of the Royal Astronomical Society* 186:327–42 (1979).

Hill, H A., and W. A. Dziembowsky, ed. *Nonradial and Nonlinear Stellar Pulsation,* Lecture Notes in Physics 125. New York: Springer-Verlag, 1980. Up-to-date articles on solar oscillations by T. P. Caudell, J. Knapp, H. A. Hill, and J. D. Logan; Knapp, Hill, and Caudell; and D. O. Gough. Proceedings of the 1979 Tucson symposium.

Hill, H. A., and R. T. Stebbins. "Recent Solar Oblateness Observations..." *Annals of the New York Academy of Sciences* (Proceedings of the 7th Texas Symposium, December 16–20, 1974) 262:472–80 (1975).

Scherrer, P. H., J. M. Wilcox, A. B. Severny, V. A. Kotov, and T. T. Tsap. "Further Evidence of Solar Oscillations with a Period of 160 Minutes." *Astrophysical Journal* 237:L97–98 (1980).

Severny, A. B., V. A. Kotov, and T. T. Tsap. "Observations of Solar Pulsations." *Nature* 259:87–89 (1976).

Stebbins, Robin. "Extended Observations of Solar Oscillations." Unpublished paper, 1980.

Chapter 6. THE SHRINKING SUN

Dunham, David W., et al. "Observations of a Probable Change in the Solar Radius Between 1715 and 1979." *Science* 210:1243–44 (1980).

Eddy, J. A., and Aram A. Boornazian. "Secular Decrease in the Solar Diameter, 1836–1953." *American Astronomical Society Bulletin* 11:437 (1979).

Helmholtz, H. L. F. "Observations on the Sun's Store of Force." (Reprint of 1908 paper.) In *Early Solar Physics*, by A. J. Meadows, New York: Pergamon Press, 1970.

"Is the Sun Shrinking? Two Views." *Science News,* 115:420 (1979).

Parkinson, John. "What's Wrong with the Sun?" *New Scientist,* April 24, 1980, pp. 201–04.

Parkinson, John H., Leslie V. Morrison, and F. Richard Stephenson. "The Constancy of the Solar Diameter Over the Past 250 Years." *Nature* 288:548–51 (1980).

Shapiro, Irwin I. "Is the Sun Shrinking?" *Science* 208:51–53 (1980).

Sofia, S., J. O'Keefe, J. R. Lesh, and A. S. Endal. "Solar Constant: Constraints on Possible Variations Derived from Solar Diameter Measurements." *Science* 204:1306–07 (1979).

Chapter 7. IS THE SUN A VARIABLE STAR?

Abbot, Charles Greeley. *The Sun and the Welfare of Man.* Washington, D.C.: Smithsonian Institution, 1929.

Foukal, Peter. "Does the Sun's Luminosity Vary?" *Sky and Telescope,* February 1980, pp. 111–14.

Gough, Douglas. "Climate and Variability in the Solar Constant." *Nature* 288: 639–40 (1980).

Hickey, J., et al. "Comments on Solar Constant Measurements from Nimbus 6 and 7." *EOS* 61:355 (1980).

Hickey, J. R., et al. "Initial Solar Irradiance Determinations from Nimbus 7 Cavity Radiometer Measurements." *Science* 208:281–82 (1980).

Moore, Patrick. *A Guide to the Stars.* New York: W. W. Norton, 1960.

White, Oran R., ed. *The Solar Output and Its Variation.* Boulder: Colorado Associated University Press, 1977. Especially articles by Claus Fröhlich; R. C. Willson and J. R. Hickey; and J. A. Eddy.

Willson, R. C., C. H. Duncan, and J. Geist. "Direct Measurement of Solar Luminosity Variation." *Science* 207:177–79 (1980).

Willson, R. C., S. Gulkis, M. Janssen, H. S. Hudson, and G. A. Chapman. "Observations of Solar Irradiance Variability." *Science*, 211:700–702 (1981).

Chapter 8. QUEST FOR THE CLIMATE CONNECTION

Eddy, John A. "Historical and Arboreal Evidence for a Changing Sun." In *The New Solar Physics,* ed. John A. Eddy. Boulder, Colo.: Westview Press, 1978.

Eddy, John A. "Climate and the Changing Sun." In *Encyclopaedia Britannica Yearbook of Science and the Future,* 1979.

Ellison, M. A. *The Sun and Its Influence.* London: Routledge and Kegan, 1959.

Herman, John R., and Richard A. Goldberg. *Sun, Weather, and Climate.* NASA SP-426. Washington, D.C.: U.S. Government Printing Office, 1978.

Lockyer, Sir Norman. *The Sun's Place in Nature.* New York: Macmillan and Co., 1897.

McCormac, Billy M., and Thomas A. Seliga, eds. *Solar-Terrestrial Influences on Weather and Climate.* Boston: D. Reidel, 1979.

Olson, Roger H. "Sun-Weather Effects." *Nature* 275:368–69 (1978).

Pittock, A. Barrie. "Solar Cycles and the Weather: Successful Experiments in Autosuggestion?" In *Solar-Terrestrial Influences on Weather and Climate,* eds. Billy M. McCormac and Thomas A. Seliga. Boston: D. Reidel, 1979.

Roberts, Walter Orr. "Introductory Review of Solar-Terrestrial Weather and Climate Relationships." In *Solar-Terrestrial Influences on Weather and Climate,* eds. Billy M. McCormac and Thomas A. Seliga. Boston: D. Reidel, 1979.

Roberts, Walter Orr. "Variations in the Sun and Their Effects on Weather and Climate." (Penrose Memorial Lecture) *Proceedings of the American Philosophical Society* 123:151–59 (1979). Informal survey with personal recollections.

Schneider, Stephen H. *The Genesis Strategy: Climate and Global Survival.* New York: Plenum, 1976.

Shapiro, Ralph. Review of *Sun, Weather, and Climate* (J. R. Herman and R. A. Goldberg). *EOS* 61:6 (1980).

Stuiver, Minze. "Solar Variability and Climatic Change During the Current Millennium." *Nature* 286:868–71 (1980).

Young, C. A. *The Sun.* New York: D. Appleton and Co., 1895.

Chapter 9. DANCE OF THE ORBITS

Beaty, Chester B. "The Causes of Glaciation." *American Scientist* 66:452–59 (1978).
Berger, A. L. "Long-Term Variations of the Earth's Orbital Elements." *Celestial Mechanics* 15:53–74 (1977).
Berger, A. L. "Support for the Astronomical Theory of Climatic Change." *Nature* 269:44–45 (1977).
Broecker, Wallace S., and Jan van Donk. "Insolation Changes, Ice Volumes, and the 0–18 in Deep-Sea Cores." *Reviews of Geophysics and Space Physics* 8:169–97 (1970).
Calder, Nigel. "The Cause of the Ice Ages." *New Scientist,* December 9, 1976, pp. 576–78.
Goreau, Thomas J. "Frequency Sensitivity of the Deep-Sea Climatic Record." *Nature* 287:620–22 (1980).
Hays, James D. "Climatic Change and the Possible Influence of Variations of Solar Input." In *The Solar Output and Its Variation.* Boulder, Colo.: Colorado Associated University Press, 1977.
Hays, J. D., John Imbrie, and N. J. Shackleton. "Variations in the Earth's Orbit: Pacemaker of the Ice Ages." *Science* 194:1121–32 (1976).
Imbrie, John, and Katherine Palmer Imbrie. *Ice Ages: Solving the Mystery.* Short Hills, N.J.: Enslow Publishers, 1979. Thorough, well-written treatment of quest to prove astronomical theory of ice ages.
Vernekar, Anandu D. "Variations in the Insolation Caused by Changes in Orbital Elements of the Earth." In *The Solar Output and Its Variation,* ed. O. R. White. Boulder, Colo.: Colorado Associated University Press, 1977

Chapter 10. CYCLES OF DROUGHT

Darrah, William Culp. *Powell of the Colorado.* Princeton, N.J.: Princeton University Press, 1969.
Dicke, R. H. "The Clock Inside the Sun." *New Scientist,* July 5, 1979, pp. 12–14.
Dicke, R. H. "Is There a Chronometer Hidden Deep in the Sun?" *Nature* 276:676–78 (1978).
Dicke, R. H. "Solar Luminosity and the Sunspot Cycle." *Nature* 280:24–27 (1979).
Epstein, Samuel and Crayton J. Yapp. "Climatic Implications of the D/H Ratio of Hydrogen in C-H Groups in Tree Cellulose." *Earth and Planetary Science Letters* 30:252–61 (1976).
Mitchell, J. Murray, Jr., Charles W. Stockton, and David M. Meko. "Evidence of a 22-Year Rhythm of Drought in the Western United States Related to the Hale Solar Cycle Since the 17th Century." In *Solar-Terrestrial Influences on Weather and Climate*, eds. Billy M. McCormac and Thomas Seliga. Boston: D. Reidel, 1979.

Chapter 11. SECTORS, STREAMS, AND STORMS

Hines, C. O. "Cause-Effect Inferences in Geophysical Statistical Studies." In *Physics of Solar Planetary Environments,* ed. Donald J. Williams. Washington, D.C.: American Geophysical Union, 1976.

Hines, C. O., and I. Halevy. "On the Reality and Nature of a Certain Sun-Weather Correlation." *Journal of the Atmospheric Sciences* 34: 382–404 (1977).

Hundhausen, Arthur J. "Streams, Sectors, and Solar Magnetism." In *The New Solar Physics,* ed. John A. Eddy. Boulder, Colo.: Westview Press, 1978

Olson, Roger H. "The Crisis in Sun-Weather Research." Unpublished manuscript, 1979.

Pittock, A. B. "Enigmatic Variations." *Nature* 283:605–06 (1980).

Roberts, Walter Orr. "Variations in the Sun and their Effects on Weather and Climate." *Proceedings of the American Philosophical Society* 123: 151–59 (1979).

Scherrer, Philip H. "Solar Variability and Terrestrial Weather." SUIPR Report No. 786, Institute for Plasma Research, Stanford University (1979).

Shapiro, Ralph. "An Examination of Certain Proposed Sun-Weather Connections." *Journal of the Atmospheric Sciences* 36:1105–16 (1979).

Siscoe, George L. "Solar-Terrestrial Influences on Weather and Climate" *Nature* 276:348–52 (1978).

Svalgaard, Leif, and John M. Wilcox. "A View of Solar Magnetic Fields, the Solar Corona, and the Solar Wind in Three Dimensions." *Annual Review of Astronomy and Astrophysics* 16:429–43 (1978).

Wilcox, John M. "Influence of the Solar Magnetic Field on Tropospheric Circulation." In *Solar-Terrestrial Influences on Weather and Climate,* eds. B. M. McCormac and T. A. Seliga. Boston: D. Reidel, 1979.

Wilcox, John M. "Solar Structure and Terrestrial Weather." *Science* 192:745–48 (1976).

Wilcox, John M. "Solar/Terrestrial Meteorological Relationships." SUIPR Report No. 719, Institute for Plasma Research, Stanford University, January 1979.

Wilcox, J. M., J. T. Hoeksema, and P. H. Scherrer. "Origin of the Warped Heliospheric Current Sheet." *Science* 209:603–05 (1980).

Wilcox, John M., and Norman F. Ness. "Quasi-Stationary Corotating Structure in the Interplanetary Medium." *Journal of Geophysical Research* 70:5793–5805 (1965).

Wilcox, J. M., and P. H. Scherrer. "Variation with Time of a Sun-Weather Effect" (Reply by R. Gareth Williams). *Nature* 280:845–46 (1979).

Wilcox, John M., Philip H. Scherrer, Leif Svalgaard, Walter Orr Roberts, and Roger H. Olson. "Solar Magnetic Sector Structure: Relation to Circulation of the Earth's Atmosphere." *Science* 180:185–86 (1973).

Wilcox, John M., Philip H. Scherrer, and Leif Svalgaard. "Intensity of Tropospheric Circulation Associated with Solar Magnetic Sector Boundary Transits." *Journal of Atmospheric and Terrestrial Physics* 41:657–59 (1979).

Wilcox, John M., Leif Svalgaard, and Philip H. Scherrer. "On the Reality of a Sun-Weather Effect." *Journal of the Atmospheric Sciences* 33: 1113–16 (1976).

Wilcox, J. M., et al. "Interplanetary Magnetic Field Polarity and the Size of Low-Pressure Troughs Near 180°W Longitude." *Science* 204:60–62 (1979)

Williams, R. Gareth, and E. J. Gerety. "Does the Troposphere Respond to Day-to-Day Changes in Solar Magnetic Field?" *Nature* 275:200–01 (1978).

Chapter 12. SEARCH FOR A MECHANISM

Markson, Ralph. "Atmospheric Electricity and the Sun-Weather Problem." In *Solar-Terrestrial Influences on Weather and Climate,* eds. Billy M. McCormac and Thomas A. Seliga. Boston: D. Reidel, 1979.
Markson, Ralph. "Solar Modulation of Atmospheric Electrification and Possible Implications for the Sun-Weather Relationship." *Nature* 273:103–09 (1978).
Markson, Ralph, and Michael Muir. "Solar Wind Control of the Earth's Electric Field." *Science* 208:979–90 (1980).
Reiter, Reinhold. "Influences of Solar Activity on the Electric Potential Between the Ionosphere and the Earth." In *Solar-Terrestrial Influences on Weather and Climate,* eds. B. M. McCormac and T. A. Seliga. Boston: D. Reidel, 1979.

Chapter 13. THE DECADE OF THE SUN

Bartusiak, Marcia F. "Experimental Relativity: Its Day in the Sun." *Science News* 116:140–42 (1979).
Chapman, Robert W. "NASA's Search for the Solar Connection—II." *Sky and Telescope,* September 1979, pp. 223–27.
Cauffman, D. P. "Solar-Terrestrial Physics in the 1980s." *EOS* 60:1014–15 (1979).
Frazier, Kendrick. "The Sun's Back Pages." *Mosaic* (National Science Foundation), 12(3):2–8, May–June 1981.
"Ganging Up on the Sun." *Science News* 117:404 (1980).
O'Leary, Brian. "The Stormy Sun." *Sky and Telescope,* September 1980, pp. 199–201.
Olson, Suzanne. "Solar Tracks in the Snow." *Science News* 118:313 (1980).
Raisbeck, G. M. and F. Yiou. "Be-10 as a Potential Probe of Solar Variability Influence on Climate." Paper presented at the International Conference on Sun and Climate, Toulouse, France, September 30–October 3, 1980.
Raisbeck, G. M., and F. Yiou. "Be-10 in Polar Ice Cores as a Record of Solar Activity." In *The Ancient Sun: Fossil Record in Earth, Moon, and Meteorites,* eds. R. O. Pepin, J. A. Eddy, and R. B. Merrill. Elmsford, N.Y.: Pergamon, 1980.
Rust, David M. "Warming Up for the Solar Maximum Year." *Sky and Telescope,* October 1979, pp. 315–18.
Space Science Board. *Solar System Space Physics in the 1980s.* Washington, D.C.: National Academy of Sciences, 1980.
Thomsen, Dietrick E. "A Flair for the Sun." *Science News* 118:152–54 (1980).
Williams, G. E. "Sunspots Periods in the Late Precambrian Glacial Climate and Solar-Planetary Relations." *Nature* 291:624–28 (1981).

INDEX